Job Motivation in Libraries

About the Authors

Dr. A. Naseer currently serves as Assistant Librarian at the Kerala Agricultural University, College of Agriculture, Vellayani, Thiruvananthapuram, Kerala. He did his LLB and MLISc from the University of Kerala, Thiruvananthapuram. He secured a Ph.D. in Library and Information Science from the University of Kerala. He is a library and information professional with more than Seventeen years of rich experience in the domain. He has co-authored a book Organizational Climate in Libraries, and published nearly 17 research papers in journals and conference proceedings. Further, he is the Best Paper Award recipient in the International Conference on CTTL-2019 held at Manonmaniam Sundaranar University, Tirunelveli, Tamilnadu. His areas of interest are management of library and information centres, human resources development, orientation and sensitization of e-resources, and intellectual property.

Dr. B. Mini Devi (Bhavani Mini Devi) is currently working as Assistant Professor and Former Head of the Department of Library and Information Science at the University of Kerala, India. She is presently the Director of the Centre for Information Literacy Studies at the University of Kerala. The University of Kerala bagged NAAC A++ Accreditation during her departmental headship. She also chaired as the Director of UIT (University Institute of Technology) of the University of Kerala. She serves as the Chairperson, Board of Studies in Library and Information Science, University of Kerala and member in University of Calicut and Kannur University. She is also a member of the Board of Examiners at the University of Kerala and Calicut University. She widely contributes to the University of Kerala through policy-making, regulation framing, and syllabus up dation as the Chairperson and Academic Council Member. She essays the role of Resource Person in Malaviya Mission Teacher Training Centre at the University of Kerala and University of Calicut.

Dr. B. Mini Devi actively participated in sensitisation and grievance redressal as an Anti-Ragging Committee Member of the University of Kerala. She also conducts innovative Conferences, (Online/Offline), Talks and Exhibitions on Library Science. She chairs as the subject rapporteur at various national and regional conferences. She also contributes as a career counsellor, subject expert, library mentor and motivational speaker at various universities and colleges. She scored the third rank in MSc. Botany, NET-JRF along with third rank in M.L.I.Sc. from the University of Kerala. She began her career with the prestigious British Library apprenticeship, Ernakulam Model Engineering College librarianship and 15 years professional experience as librarian. She completed her PhD in "Toxicology Literature: An Informetric Analysis" from Cochin University of Science and Technology, Ernakulam, Kerala, India. She specializes in library classification, cataloguing, informetrics, scientometrics, research methodology, statistics & data analysis, information literacy, media literacy, copyright & intellectual property rights, library management, KOHA, data mining, machine learning, Google SEO, AI, ChatGPT, and information storage & retrieval. She published a wide range of books and book chapters. Alongside 200 co-authored articles, she published over 150 articles in national & international journals and conference proceedings. She produced 20+ doctoral PhD thesis, 2 postdoctoral theses and 5 MPhil dissertations under her research guidance and mentorship. She believes in unbiased, student-friendly teaching through the humble educational motto of the University of Kerala—Karmani Vyajyate Prajna (Wisdom manifests into action).

Job Motivation in Libraries
Changing Perspectives

A. Naseer
Assistant Librarian
Kerala Agricultural University, College of Agriculture
Vellayani, Thiruvananthapuram, Kerala

B. Mini Devi
Assistant Professor and Former Head
Department of Library and Information Science
University of Kerala
Kerala

PMW

PMW
FU 75, Pitam Pura
New Delhi 110034
Mob. 91-8447075807
Email: publishmywork23@gmail.com
Web: www.publish-mywork.com

Print ISBN: 978-93-61348-54-9

ebook ISBN: 978-93-61348-56-3

Preface

Strategic Human Resource Management (HRM) has a prominent role in any organization, including libraries and information centres. Job motivation is employed in the modern era in many service organizations, including libraries and information centres, to improve the performance of employees and fulfil the objectives of the organizations. Motivation prompts to do something and helps willingness to perform various tasks in an organization effectively.

The present study examines the job motivation of Library and Information Science professionals working in the university libraries of Kerala. This original and comprehensive work is based on a research on the subject at the University of Kerala, Thiruvananthapuram. While presenting this topic, we have in our mind about the requirements of the students of higher learning institutions and a larger audience interested in the subject. It was felt that such a study could provide some valuable insights for planning and managing libraries and information centres over the coming years or so.

An attempt has been made to deal with the subject matter comprehensively even as we have tried to be concise. The book is presented in six chapters, and suggestions are made to improve the job motivation of library and information professionals working in university libraries.

The authors indebted to several institutions, well-wishers and friends for their liberal assistance to conduct this study. We also indebted to University of Kerala, Thiruvananthapuram, for arranging facilities and services but for which the study would not have been completed at all. Moreover, we have no words to describe our gratitude to the Library and Information Science Professionals of the universities of Kerala, who are kind enough to make valuable data for conducting this study. We also take this opportunity to express our special thanks to Publish My Work, New Delhi for bringing out this book in the most professional way.

Authors

Acknowledgements

The book is the product of ideas and experiences gathered by us over a long period of time from the LIS professionals working in the university libraries of Kerala. Sincere thanks are due to University of Kerala for giving us the opportunity for pursuing this work. We are grateful to all the LIS professionals who have given data for the preparation of this book. Finally, our gratitude goes out to all near and dear ones whose good wishes and feedback have made our academic and research endeavour truly fruitful.

Contents

List of Abbreviations

BLISc : Bachelor Degree in Library and Information Science
CAB : Centre for Agriculture and Bioscience
CAS : Current Awareness Service
CD ROM : Compact Disc Read-only Memory
CeRA : Consortium for e-Resources in Agriculture
CUSAT : Cochin University of Science and Technology
CUSATL : Cochin University of Science and Technology Library
DDC : Dewey Decimal Classification
DELNET : Developing Library Network
DOAJ : Directory of Open Access Journals.
DOAR : Directory of Open Access Repositories
HRM : Human Resource Management
IASLIC : Indian Association of Special Libraries and Information Centres
ICAR : Indian Council of Agricultural Research
ICT : Information and communication Technology
IFLA : International Federation of Library Associations Institutions
INFLIBNET : Information and Library Network
JCCC : J-Gate Custom Content for Consortia
JSTOR : Journal Storage
KAU : Kerala Agricultural University
KAULIS : Kerala Agricultural university Library and Information System
KELPRO : Kerala Library Professionals' Organisation
KSHEC : Kerala State Higher Education Council
KU : Kannur University
KUL : Kerala University Library
KAUL : Kannur University Library
KVASU : Kerala Veterinary and Animal Sciences University
KVASUL : Kerala Veterinary and Animal Science University Library
LIS : Library and Information Science
LISTA : Library, Information Science & Technology Abstracts
M.Sc. : Masters of Sciences
MGUK : Mahatma Gandhi University, Kerala

MGUL	: Mahatma Gandhi University Library
MHRD	: Ministry of Human resource development
NEP	: National Education Policy
NUALS	: National University of Advanced Legal Studies
OPAC	: Online Public Access Catalogue
Ph.D.	: Doctor of Philosophy
SDI	: Selective Dissemination of Information
SPSS	: Statistical Package for Social Sciences
SSUS	: Sree Sankaracharya University of Sanskrit
SSUSL	: Sree Sankaracharya University of Sanskrit Library
UCE	: University College of Engineering
UGC	: University Grants Commission
UIT	: University Institutes of Technology
UN	: United Nations
UoC	: University of Calicut
UoCL	: University of Calicut Library, CHMK Library
UoK	: University of Kerala
USG	: University System of Georgia
UTE	: University Teacher Education Centres
WHO	: World Health Organization
Wi Fi	: Wireless Fidelity

1

Introduction

Introduction

Human resources are the most important assets of an organization. The success of an organization depends on the talent and caliber of the employees working therein. It is true in the case of service-oriented institutions like libraries and Information centres. Libraries of the university are a treasure trove of information and knowledge. The library is one of the important components of a university. Library and Information Science (LIS) professionals are the assets of libraries and information centres. Libraries are undergoing transformational changes across the world as a result of disruptive technologies. Libraries are becoming informal, relaxed and collaborative spaces. Information and Communication Technology (ICT) has changed the role and functions of libraries. There are also changes in the nomenclature of libraries to information centres, information resource centres, green libraries and open libraries.

Along with traditional and print documents, collections of modern libraries include electronic resources and web resources. Services rendered by the libraries are moving towards online information access and retrieval. Motivating Library and Information Science (LIS) professionals in the university libraries will benefit the academic community throughout and beyond their academic careers. It has necessitated conducting a study on the job motivation of LIS professionals in the university libraries of Kerala.

Reaching out to library users at the moment is more important in the virtual learning environment. Job motivation is one of the innovative ways to reach out the users through proactive engagements. Job motivation can make LIS professionals connect with users of the library and improve the library's services. Motivating LIS professionals can empower users of the library to search for information on their own. In the modern era, job motivation is employed in many organizations, including service institutions like libraries and information centres, to fulfil the organisation's well-defined objectives and improve personnel performance. It is essential to fulfilling the vision, mission and mandate of the university libraries. Motivation is a state of the inner driving force that will strive to act in a certain way. Motivation prompts to do something and helps willingness to perform various tasks in the organization efficiently. The factors include mood and motivation of staff, which influence the organisation's performance and outcome.

This chapter presents the importance of job motivation to the LIS Professionals working in the university libraries of Kerala. The chapter includes job motivation, Higher Education, university libraries, definitions, statement of the problem, need for the present study, objectives of the study, hypotheses, methodology, scope and limitation of the study and organization of the study.

Job Motivation

The word "motivation" is derived from the word 'motive' in the English language, which is defined as a need that requires satisfaction. Motivation is an important part of human psychology. Employee motivation can be defined as "psychological forces that determine the direction of a person's behaviour in an organization, a person's level of effort and a person's level of persistence" (Jones and George, 2008). It is important to understand that employee motivation is a separate and distinct topic apart from motivation. Extensive studies exist on this topic indicating the level of importance of employee motivation in the success of organizations. Employers who understand the theories of motivation have a greater ability to understand what motivates employees, boost employee morale and thus obtain the advantages of greater organizational service and productivity of organization and performance of employees.

Job motivation is a psychological feature that induces an organism to act towards a desired goal, elicits controls, and sustains certain goal-directed behaviours. It can be considered a driving force, a psychological one that compels or reinforces an action towards a desired goal. Job is a piece of work, especially a specific task done as part of the routine of one's occupation for an agreed price.

Motivation is concerned with all the states of inner striving, often described as drives, desires or wishes that make us act in certain ways. It is that inner state that activates or moves us all. It is very pertinent to note that without motivation, there would be no purposive, organized behaviour by the people either at work or elsewhere (Yalokwu, 2006). Motivation prompts to do something and helps willingness to perform various tasks. Authorities and professionals can perform a pivotal task in the area of motivation of job. Motivation encourages a conducive working environment in the organization.

The management principles in an organization include planning, organizing, staffing, directing, coordinating, reporting and budgeting. It is widely used in the field of management. For workers to be effectively directed and perform better in their roles, staff motivation must be given adequate attention, i.e., good leadership qualities, staff training, participatory management, a conducive working environment and free flow of information, etc. (Aina, 2004). Motivation energize and stimulates to change human behaviour for the attainment of the objectives of the organization. Motivation drives personnel to the application of their talent, calibre and skills to the maximum extent. Motivation also empowers the personnel to be more committed to their duties and responsibilities, which ultimately lead to the progress of the organization.

Higher Education

From time to time, the Government of India has constituted several commissions to achieve the task of higher education in the country. The first University Education Commission was appointed in the year 1948. In 1952, a secondary education commission was constituted. To ensure all-around development in the field of education, the Ministry of Human Resource development (MHRD) was constituted in 1985. In 1986 National Policy on Education recognized the role of higher education in national development. It also suggested that State Councils of Higher Education should be set up in all states (National Policy on Education, 1986). The National Education Policy 2020 (NEP 2020) outlines the vision of India's new education system. National Educational Policy 2020 is a comprehensive framework for elementary education to higher education (National Educational Policy 2020).

Kerala State Higher Education Council

Kerala State Higher Education Council (KSHEC) came into existence in 2007. KSHEC brings together and expands the rich array of teaching, research and human resource development. It is an advisory body to the Government of Kerala in higher education and acts as a state-level coordinating agency for higher educational activities. KSHEC also bring together scholars and stakeholders in state's education to materialize its vision to revitalize the higher education sector of Kerala and foster academic inquiry and research (Kerala State Higher Education Council, 2019).

Higher Education in Kerala

Kerala has the highest proportion of literate persons among the Indian States. The effective literacy rate is 93.91 per cent as per Census 2011. In Kerala, 96.02 per cent men and 91.98 per cent of women are literate as against 82.14

per cent of men and 65.46 per cent of women at the all-India level. There are 14 State universities functioning in Kerala. Out of these, four universities, namely, University of Kerala, Mahatma Gandhi University, Kerala, University of Calicut and Kannur University, are general in nature and offer various courses. Sree Sankaracharya University of Sanskrit, Thunchath Ezhuthachan Malayalam University, Cochin University of Science and Technology, Kerala Agricultural University, Kerala Veterinary and Animal Science University, Kerala University of Health Sciences, Kerala University of Fisheries and Ocean Studies, Kerala Technological University offer specialized courses in specified subject areas. Besides these, the National University of Advanced Legal Studies (NUALS) established in 2005, and a Central University in Kasaragod District are also functioning in the State. There are 219 Arts and Science Colleges comprising 156 private aided colleges and 63 Government colleges in the State. Apart from these, various arts and science colleges, un-aided/self-financing, are also functioning with the affiliation of universities in the State. In 2017-18, Ernakulam (26) district had the largest number of arts and science colleges, followed by Kottayam (23) district. Thiruvananthapuram and Kozhikode districts have the most significant number of Government colleges (10) in the State, (State Planning Board, Kerala Government, 2018).

University Libraries

These days, people mainly depend on search engines to cull information. But there can hardly be a replacement for the library. It is said that Google can answer a thousand questions, but the librarians can give the right one. With the emergence of technology, readers are using e-books and other online resources on their cell phones and other gadgets to enhance the availability and accessibility of information and knowledge. Data and information are also obtained through social media. However, libraries still attract users to the library. University libraries are active and have a good collection of print and digital documents to fulfil the academic requirement of their user fraternity. It is a place to collect academic knowledge. Libraries are the university's knowledge centres to strengthen the university's activities in a timely manner.

University libraries are libraries attached to universities mainly for serving the academic community of the university. University library is the collective name applied to the library, established, administered and funded by universities that do the information, research and curriculum needs of the students, researchers, faculty, staff and members of the various academic bodies of the universities. The University library is established as a part of the university to support learning, teaching, research and services to the clientele and community. Ranganathan (1957) rightly pointed out that the library is a growing organism. Libraries are undergoing rapid changes as a result of digital disruption. Nowadays library is a complex organizational system. Motivating employees to perform the work assigned to them is one of the chief goals of an organization. Several factors are motivating the job of LIS professionals working in the libraries of the universities. The present study highlights the job motivation of LIS professionals working in universities in Kerala.

The library is a non-profit organization. It is a service institution and is essential in rendering service and dissemination of information to clientele. In libraries, employees experience satisfaction concerning the service they rendered. The clientele also gets the benefit when they get service from the LIS professionals. Job motivation has a managerial effect in university libraries. It is instrumental in human resource management in libraries. There are well defined hierarchical structures of the library staff in the university libraries in Kerala. Human resource management has a prominent role in keeping controlling and organizing facilities and resources of the libraries.

Definitions

Job Motivation

Job motivation is defined as the "motivation that arises from the pleasure that individuals obtain from their job and their resulting greater likelihood to engage in job-related tasks willingly" (Hackman and Lawler, 1971; Tietjen and Myers, 1998).

In the present study Job motivation refers to the motivation of LIS professionals in their work.

Library

In the present study, the library refers to the libraries attached to the state universities in Kerala, India.

Kerala

Kerala state lies between 8^0 18' and 12^0 48' North Latitude and between 74^0 52' and 77^0 24' East longitude. The modern state of Kerala was formed on 1 November 1956. The state has a total area of 38,864 square K.M. and a seacoast that is 580 K.M. long. It varies in width from about 11 K.M. to 120 K.M. (Sreedhara Menon, 1979).

Statement of the Problem

The productivity of LIS professionals in academic libraries depends on many factors. Libraries are changing organic entities that constantly adapt and become something else (Crawford,2017). The library is the heart of the university. The University library is a knowledge resource centre. It caters to the academic needs of students, teachers, research scholars and scientists. Harrod's Librarian's Glossary defines University Library as "a library or group of libraries established, maintained, and administered by a university to meet the needs of its students and members of the academic staff" (Prytherch,1995). The enrollment of students is increasing in the universities. The development of university education can be achieved through a well-equipped library. It was noted that the job motivation of LIS professionals influences academic libraries. Though libraries attached to the universities face several challenges and opportunities in the modern era as far as the dissemination of information is concerned. They also offer services to the clientele that provide information in different ways to facilitate the availability and accessibility of information. The university libraries are non-profitable entities maintained by the universities exclusively for academic purposes.

Moreover, university libraries can function as a knowledge centre where stakeholders of the universities can generate new ideas and share their views. Because of these developments, university libraries should act proactively where the academic community gathers information according to their interest and works as a resource centre for the academic community. It provides learning, teaching and research opportunities for all stakeholders of the university. It increases the effectiveness of library services. Therefore, job motivation is an excellent predictor of an organization as well as employee performance. Thus, the present study attempts to probe into these aspects relating to job motivation among the library and information professionals in universities in Kerala.

Need for the Present Study

In recent years, considerable research has been done in the field of job motivation. Job motivation is nowadays considered a global concept. The majority of the studies are conducted in business organizations. However, studies are also found in the areas of Library and Information Science. Drastic changes in technologies and integration of several innovating technologies are changed libraries and information centres worldwide. The library is the heart and soul of university education. Library-oriented education has attained much attention in higher education. The libraries are introducing and providing various new services to the clientele. To meet the clientele's ever-changing demands, the LIS professionals are required to motivate for the up-gradation of knowledge and effective dissemination of information. In this scenario, it is very relevant to conduct a study like this to measure the motivational factors of LIS professionals. Motivation has a direct impact on the services provided by library professionals. The study aims to improve services provided by the LIS professionals of the University Libraries in Kerala and improve job motivation. In Kerala, the number of universities is increased after the formation of state. At present, libraries in Kerala are not far behind in keeping pace with time. Services provided to the clientele have to be increased by utilizing all recent technologies. The present study is an attempt to investigate the job motivation of LIS professionals in the university libraries in Kerala. The study's outcome will be helpful for LIS professionals and authorities of the universities to improve the current service set up of the university libraries. The finding and suggestions from the study are useful for library management to create motivation for the job for LIS professionals. Hence, the use of the library and its resources along with the productivity and well-being of LIS professionals may be enhanced accordingly.

Objectives of the Study

The study's primary objective is to determine the job motivation of LIS professionals working in the university libraries of Kerala. The following are the important objectives of the study.

- To identify the factors that influence the LIS professional's job motivation working the university libraries of Kerala.
- To study the Job motivation of the LIS professionals working in the university libraries of Kerala.
- To find out the total motivation score of all the factors selected for studying job motivation.
- To rank the factors which are affecting the job motivation of LIS professionals in the universities of Kerala.
- To examine the role of authorities in motivating the job of LIS professionals of university libraries in Kerala.
- To study the role of Library Associations in motivating the job amongst qualified LIS professionals.
- To suggest measures for enhancing job motivation in the university libraries of Kerala.

Hypotheses

H1: There is a lack of motivation among the LIS professionals to carry out their duties and responsibilities.

H2: Authorities are adequately motivating the job of the LIS professionals in the universities of Kerala.

H3: The role of Library Associations is minimum in motivating the job of qualified LIS Professionals.

Universe and Study Design

As per the official records, there are fourteen state universities in the state of Kerala in 2021. All the universities have their act and statutes. The state universities have their libraries. The present study was conducted in the selected eight state universities of Kerala. There is 385 sanctioned post of LIS professionals in the selected eight universities. In the present study, the total number of LIS professionals working in the universities is 276. The post of the University Librarian in all the universities are lying vacant. The cadre of LIS professionals in the universities is Deputy Librarian, Assistant Librarian, Reference Assistant (Higher Grade)/Reference Officer (Higher Grade), Junior Librarian/Reference Assistant/Reference Officer, Technical Assistant/ Professional Assistant Grade. I and Library Assistant/Professional Assistant Grade. II. All these are the sanctioned cadre in the libraries. The entire cadre of LIS professionals is selected for the study. The professionals with minimum professional qualification of Bachelor Degree in Library and Information Science (BLISc), are considered for the study.

Population and Size

For the present study, the census method was used for the collection of data. Data gathered through both open-ended and closed-ended questionnaires. A structured questionnaire was designed and distributed to all the 276 LIS professionals working in the selected eight universities. Out of it, 266 questionnaires were duly filled and returned. The response rate was 96.38 %. The study is mainly based on the primary data obtained from the 266 LIS professionals.

Scope of the Study

The present study is confined to the analysis of job motivation among the LIS professionals working in Kerala's eight major state universities. The libraries of the universities are:

1. University of Kerala, Thiruvananthapuram
2. University of Calicut, Malappuram
3. Mahatma Gandhi University, Kerala, Kottayam

4. Cochin University of Science and Technology, Kochi
5. Sree Sankaracharya University of Sanskrit, Ernakulam
6. Kerala Agricultural University, Thrissur
7. Kannur University, Kannur
8. Kerala Veterinary and Animal Science University, Wayanad

Methodology

In terms of the objectives, both primary and secondary sources of information have been consulted to obtain relevant data in the present study. The primary data of the present study is collected from the LIS professionals working in the universities. For collecting primary data, a well-structured questionnaire has been designed to elicit data from the LIS professionals. Moreover, the researcher himself has experience and knowledge about the functioning of the library system, various aspects related to the job of LIS professionals. Observation and interaction with fellow professionals have also helped to ascertain requisite information for the study. Secondary data for the analysis is collected from the records of universities and the government.

Literature Search

A literature search was also done among the primary, secondary and tertiary sources of information.

Tool used for Data Collection

A structured questionnaire was designed as a data collection tool and circulated among LIS professionals keeping because of the specific objectives to undertake the present study. The questionnaire is divided into two parts. The first part intended to elicit general information from the LI professionals. The second part of the questionnaire is to elicit the opinion of LIS professionals regarding Job motivation in the libraries of the universities. The Likert 5-point scale is used for investigating the job motivation of LIS professionals in libraries. The questionnaire was used to collect the data from the LIS professionals. Secondary data were collected from guide books, encyclopedias, reference books, brochures, annual reports, abstracts, directories and websites of the university libraries and universities. The collected data were examined and analyzed.

Period of Study

The present study is confined to the period from 2014 to 2019. On the other hand, the data related to the study are collected from October 2017 to March 2018.

Tools used for the analysis of Data

The data collected for the study were processed and analysed statistically using appropriate description methods to ensure data reliability and appropriateness of solutions to the problems. Statistical Package for Social Sciences (SPSS. Version 23) and the latest version of Micro-Soft Excel was used for this purpose. Mathematical and statistical techniques such as descriptive statistics, 't' test is used to elicit information.

Limitations of the Study

The study has certain limitations. The study covers only the job motivation of LIS professionals in the university libraries in Kerala. The study is limited to the professionally qualified and permanent LIS professionals working in the eight-state universities of Kerala. Universities established recently, and libraries with skeletal staff, library and information professionals working on a contract basis, and daily wage basis are exempted from the study's purview.

Organization of the Study

The entire theme of the present study is analysed and presented in the following six chapters.

Chapter - 1 Introduction

The chapter presents the introduction, definitions, objectives of the study, hypotheses formulated, the need of the present study, the methodology adopted for the study, scope and limitation of the study.

Chapter – 2 Review of Literature

It makes a view on the literature related to various studies on job motivation.

Chapter – 3 Methodology and Profile of the Universities

The chapter deals with the methodology adopted for conducting the present investigation, along with the profile of the universities selected for the study.

Chapter – 4 Job Motivation in University Libraries: An analysis

The chapter deals with a bird's eye view on job motivation and various theories associated with job motivation.

Chapter – 5 Analysis and Data Interpretation

The chapter presents an analysis of various aspects relating to job motivation.

Chapter – 6 Findings and Conclusion

The findings and conclusion of the study have been presented followed by a selected bibliography.

References

Aina, L. O. (2004). Library and Information Science Text for Africa (pp.281-282). Ibadan: Third World Information Service Ltd.

Crawford, A. (2017). The Meaning of the Library: A Cultural History. Princeton: Princeton University Press.

Hackman, J.R., & Lawler, E.E. III. (1971). Employee reactions to job characteristics, Journal of Applied Psychology, 55 (3), 259-286.

India, Ministry of Human Resource Development (2020). National Educational Policy 2020, Government of India.

India, Ministry of Human Resource Development. (1986). National policy on Education-1986(p.15). New Delhi: Department of Education.

Jones, G. R., & George, J. M. (2008). Contemporary management. Boston: McGraw Hill/Irwin

Kerala State Higher Education Council (2019). Official website of Kerala State Higher Education Council. Retrieved 2019, from the official website of Kerala State Higher Education council. http://www.kshec.kerala.gov.in.

Prytherch, Ray. (1995). University Library. In Harrod's Librarians' Glossary (8th ed., p.663). England: Gower.

Ranganathan, S R. (1957). The Five Laws of Library Science (2nd ed.). Bombay: Asia Publishing House.

Sreedhara Menon, A. (1979). Social and Cultural History of Kerala (pp.3-4). New Delhi: Sterling Publishers Pvt. Ltd.

State Planning Board. (2018). Economic Review, 2018. Thiruvananthapuram, Kerala, India. Retrieved from http://www.spb.kerala.gov.in

Yalokwu, P.O. (2006). Fundamentals of Management (2nd ed., p.396) Lagos: African Centre for Management and Education.

Tietjen, M.A., & Myers, R.M. (1998). Motivation and job satisfaction, Management Decision, 36 (4), 226-231.

2

Review of Literature

Introduction

This chapter reviews literature related to the area of the current study. Scrutiny of the available literature is of great value in gaining insight into the research problem under investigation. As the studies on the job motivation in the university libraries in Kerala is less, the works reported other places were reviewed that are relevant to the different areas of present research. Hence, the available studies related to the topic of research are exhaustively reviewed. In order to survey the previous studies on the topic, a number of primary, secondary and tertiary sources were consulted. The reviews are arranged under the following headings.

- **Job Motivation in General**
- **Job Motivation in Libraries**

Job Motivation in General

Asante., Yahaya and Yaahowaa (2020) attempted to examine the success and challenges of Ghana's Single Spine Pay Policy. One hundred and sixty civil and public servants were chosen from Tamale Metropolis using a simple random sampling technique. Data were analysed to investigate relationships and predict outcomes. The study revealed that seventy per cent of the civil and public servants are not satisfied with the policy. It was recommended in the study is that seminars and workshops should be organized to enlighten the concepts of single spine salary structure.

Raj and Devakumar (2018) assessed the relationship between job motivation and job involvement. The study was conducted among the employees of general co-operative marketing societies in Kerala. Both primary data and secondary data were collected from 385 employees. The study found that there is a significant relationship between motivation and job involvement of the employees.

Sadeghi and Sattari (2018) compared job motivation and work engagement. The study population is all the teachers, in the academic year 2016-17, working in smart and ordinary middle and high schools in Parsabad, Ardabil province, Iran. Morgan table was used to select teaches by a stratified random sampling method. The result of the study revealed that there was a significant difference between teachers working in smart and ordinary middle and high schools in terms of job motivation and work engagement than those working in ordinary schools. The result also indicated that the job motivation and work engagement components were higher among teachers of smart schools.

Gairola (2015) assessed the job performance and preferences for various streams of activities within the establishment of forest guards in the Indian states of Maharashtra, Karnataka and Andhra Pradesh. The study ascertained how the predictor variables of job motivation, job satisfaction, job stress, age, educational qualification, length of service, gender and body mass index influence job performance. The study also explored the human resource management issues related to forest guards. The study suggests that job motivation, job satisfaction, and educational qualification positively influence job performance. In the study, job stress, age, and length of service have a negative relationship with job performance. Moreover, gender and body mass index have no relationship with job performance.

Mohammed et al. (2015) conducted a cross-sectional study to validate job motivation in Gamo Gofa Zone, Southern Ethiopia. The study was conducted in 150 health care posts in Gamo Gofa Zone. The questionnaire was used to collect data from 301 participants. The questions are relating to socio-demographic characteristics and job motivation. The study results indicate that eight factors emerged during exploratory factor analysis from the three dimensions of the job motivation scale: education career, workload, financial incentive, supervisor encouragement, community recognition, access to infrastructure, living conditions, and better achievement in work.

Popoola (2007) surveyed how the workplace, biographical and motivational factors affect organizational commitment of records officers in federal universities in Nigeria. In the study, there was a significant difference in organizational commitment of the respondents on their marital status, work motivation and job tenure but not on their place of work and religious work. The study recommended that records officers should be adequately motivated, and qualified manpower should be recruited to serve in the records offices to enhance their organizational commitment and work productivity.

Job Motivation in Libraries

Omol (2020) investigated the relationship among job satisfaction, leadership styles and career indecision of library personnel. The study was conducted in private universities in South-West, Nigeria. The population of the study consists of one hundred and five library personnel from five selected libraries. The questionnaire method was used as a tool for data collection. In the study, there was a significant relationship between job satisfaction and career indecision. The study also revealed that there was no significant relationship between leadership style and career indecision. It was also pointed out that there was a significant relationship between job satisfaction and leadership style towards predicting career indecision of library personnel.

A joint study by Alajmi and Alasousi (2018) examined levels of motivation and needs satisfaction of academic library employees. The study adopted Maslow's hierarchy of needs to examine individuals' motivation. The study collected and analyzed quantitative survey data from 108 employees working across eight college libraries at Kuwait University. In the study, the employees agreed that their needs were being satisfied at each of the five levels of Maslow's hierarchy and reported higher satisfaction levels of their self-actualization and social needs. Self-actualization is the summit of Maslow's motivation theory. Self-actualization represents the quest for reaching one's full potential as a person.

Bamgbose and Ladipo (2017) investigated the influence of motivation on employees performance and productivity in some academic libraries in Lagos State, Nigeria. The study adopted a survey research design with a sample of 322 library staff from four selected academic libraries. Three hundred and twenty-two copies of the questionnaire were administered on library staff, out of which 266 representing 80.0 per cent were returned. The study found that job security, wages and salary, relationship with colleagues, staff appraisal, financial incentives, and rewards were available to library employees. Most of the motivational parameters influence the performance of the library employees to a very great extent. The study showed that many staff was uncertain about the future of their career. But there was also a lack of complete senior management commitment and support. The study suggests that employees in academic libraries should be well and adequately motivated from time to time to boost their morale for efficiency and higher productivity.

Sharma and Ashish (2017) examined job involvement as a component of work culture prevailing among the library professionals of 54 special libraries. The study is based on a survey conducted on 280 library professionals of special libraries in and around Delhi and the National Capital Region. The results indicate no significant difference concerning the job involvement of professionals of all ranks and files of these 54 special libraries. The study observed that the overall job involvement of these library professionals is not very promising and measured around a moderate level.

Nazim (2016) attempted to understand the attitude of library professionals to share knowledge and identify barriers that hinder knowledge sharing among library professionals in Aligarh Muslim University. The study reveals that library professionals' positive attitudes towards knowledge sharing, share knowledge for finding solutions to the job-related problem and improving job performance. The most used information technology-based tools for sharing knowledge are e-mail, mobile, telephone, social media, blogs, etc. They acknowledged the importance of intrinsic motivation in knowledge sharing rather than extrinsic motivation. Lack of communication, lack of time and training, political and cultural reasons are major barriers perceived to knowledge sharing in libraries.

Rajpurohit (2016) explored job satisfaction among library professionals in medical, Ayurveda and dental libraries, college and university libraries in Rajasthan. The social recognition of the library professionals has assessed by consideration of aspects such as respect and recognition, status, privileges, awards and intra-personnel relationships. The study finds that job satisfaction contributes significantly to employee productivity and morale. The organizations which take care to develop individual attitudes among employees, which contribute to job satisfaction, will be benefited substantially.

Irfan., Haneefa and Shyni (2015) investigated workplace learning. The study was conducted among the library professionals in the university libraries of Kerala. The study shows that majority of the library professionals are engaged in workplace learning. The study found that formal learning methods contribute to workplace learning in university libraries. The library professionals engage in workplace learning keep them up-to-date. The study also found that most professionals are self-motivated and acquire the skills for digital library development through learning interventions. The study reports that a positive attitude with self-motivation is the major motivator for workplace learning.

Ali Al-Aufi (2014) assessed employees' status and level of motivation working in the Omani academic libraries at Muscat based on Maslow's hierarchy of needs. The study described motivation from the perspectives of librarians and other employees. The study pointed out that the motivation level was modest with varied attitudes for individual motivational needs. The need for security was indicated as the least motivating. Their lower-order motivational needs are satisfied except for security needs which demonstrated a level below satisfaction. On the other hand, the upper-level needs of self-esteem were not adequately satisfied.

Kont and Jantson (2014) examined salary conditions among the staff of Estonian university libraries. The study results are analyzed based on long-term working experience in Estonian academic libraries in light of the literature, the legislation of Estonia, and the authors' opinions. The study found that the biggest challenge for Estonian university libraries is salary. Estonian university libraries increase employees' salaries and thereby improve the economic status of employees. The study highlighted that the librarians would be well motivated for greater performance and greater productivity if their material needs were met extensively.

Madukoma et al. (2014) conducted a study on the influence of training and motivation on the job performance of library professionals at the University of Lagos in Nigeria. The study noticed that library users frequently complained that library staff does not attend to their queries properly. This could be attributed to a lack of training and motivation for the staff. The result of the study showed that training motivates library personnel to do their jobs. In turn, training enhances the performance of library personnel on their job. The study recommends that regular training would increase productivity and make available skilled workforce and personal growth.

Amune (2013) examined job motivation as a predictor of job satisfaction among professional and non-professional library staff in Ambrose Ali University, Ekpoma. The study revealed that there are motivational packages that significantly predict the satisfaction of library staff. The study also showed no significant difference between the satisfaction derived by professional and non-professional library staff. The library staff derived most satisfaction from salary or pay, library policies and administration, advancement and personal growth and job security. Therefore, the management of the library should intensify and sustain policies that will enhance the factors. Since needs are not static, top management should ensure regular monitoring of staff satisfaction in the institutions' library sub-unit to provide a well-satisfied workforce that can discharge better services to library users.

Babalola and Nwalo (2013) studied job motivation on librarians' productivity and examined the influence of job motivation on the productivity of each librarian in colleges of education, Nigeria. The study revealed that the majority of the sampled librarians is not productive. Besides, job motivation influenced the productivity of librarians to a significant extent. The study describes a need for collaboration between college authorities and librarians to optimize the productivity of the librarians. Such collaboration would include efforts made to boost the job morale of the librarians.

Henczel and Macauley (2013) explored the roles and responsibilities of professional associations in developing, maintaining and sustaining professionalism that underpins the profession of librarianship. The study discussed the roles and responsibilities of library associations as the profession evolves within its changing environment. The study intends to start a debate on what constitutes professionalism today as we consider how well society views and accepts the profession as a whole. Professionalism affects societal consideration and professional identity, with each of those influencing the profession's future and what that future may look like.

Lamptey et al. (2013) discussed the effort of public university managements to motivate librarians in public universities in Ghana. Motivation, effect of motivation of performance, views on motivation and level of satisfaction of librarians were analyzed in the study. The public university managements have made some efforts to motivate librarians. But the, Public university managements need to do more to improve the current situation. The study concludes how public university libraries can enhance motivational systems. The study provides insights into both intrinsic and extrinsic factors that motivate librarians. The study will be of great value to managers in the public university libraries in getting the best out of their staff if adequately motivated.

Owolabi et al. (2013) gathered information on librarians' perception towards monetary and non-monetary incentives in university libraries in Nigeria. The findings revealed that librarians are aware of both monetary and non-monetary incentives and that majority of the librarians benefited from monetary incentives. Motivation, job satisfaction, and increased organizational commitment are some of the benefits librarians derive from monetary and non-monetary incentives. Recommendations of the study included linking reward directly to performance and the need to pay reasonable salaries and wages to librarians to make them more effective.

Keogh (2013) examined motivation for grant writing among academic librarians. Librarians identified as successful grant authors were sent invitations to an online survey. The data collected were analyzed in the context of existing studies of librarian motivation, especially those relating to academic publishing and entrepreneurship, and the theoretical frameworks of motivation, including those of Maslow and Herzberg.

Since 2004 the Niger Delta University Library has adopted a job rotation program to help solve manpower and skills acquisition problems. Baro (2012) evaluated the Niger Delta University Library job rotation program, including its impact on newly employed librarians and their job performance, and the university library. The study revealed that the job rotation program is an important program for allowing librarians to acquire new skills, enhance staff productivity, develop new relationships across the university library and gain skills needed for future career advancement. The participants in the job rotation program who were interviewed showed a high level of satisfaction with the program. The findings will inform library administrators or management about the adoption of job rotation programs in their libraries.

Inceoglu et al. (2012) examined age-related differences in work motivation in two samples of individuals who completed a comprehensive motivation questionnaire for selection or development purposes. In one sample, age differences were examined by controlling for gender and investigating whether relationships between age and motivation were non-linear. Statistically significant relationships between motivation and age were found for most motivation scales explaining up to 12 per cent of the variance in specific scales. The other sample was used to confirm these results and determine whether differences in these motivation scales could be explained by additional demographic variables, which were not available in the first sample. When controlling for demographic variables such as gender, managerial experience and university education, the pattern of results was similar in the second data set, although effects were smaller. The study results suggested a shift in people's motives rather than a general decline in motivation with age. Older employees were less motivated extrinsically but more intrinsically in rewarding job features.

Thomas and Holley (2012) examined ways to motivate staff whose job assignments include repetitive library tasks. The study found ways to increase employee motivation, including enriching the work experience, effective communication, appropriate incentives, and treating each employee as an individual. Sharing tasks among staff and maintaining an ergonomic workspace is also important. Library automation and eliminating unneeded tasks can reduce repetitive tasks but can also lead to a reduction in staff. The study provides practical steps for managers to keep staff motivated when faced with repetitive tasks and summarize the management literature on this topic to provide practical guidance for library managers.

Farler and Preston (2012) analyzed workplace stress in a further education college library service. The study's findings are that librarians' interaction with students can be stressful or enjoyable, depending on context. The need to control noise levels, modify student behaviour and balance the needs of different user groups are cited as stressors. The result showed that library staff exhibit a degree of humour, self-awareness in their work and employ a range of methods to cope with stress, and positive stress may motivate librarians to engage actively with students and thus create job satisfaction. Negative stress may be managed by measures such as zoning and flexible governance. The study contributes to knowledge and understanding of stress in the library profession by contributing a case study of stress in the college sector.

Yahya et al. (2012) focused on improving job design for library assistants at UTHM. The study was conducted due to UTHM library assistants' complaints regarding dissatisfaction and discomfort on their job design. As the result, a new job design was produced so as to improve productivity of library assistants' satisfaction.

Bernstein (2011) discussed job satisfaction among University System of Georgia (USG) academic libraries relative to organizational classification status. The research concluded that confirmation of Frederick Herzberg's motivation-hygiene theory and USG librarians are classified as faculty at their respective institutions. A great percentage of USG librarians regarded such status as secondary to their overall job satisfaction. Intrinsic enjoyment of their profession and sense of involvement with an educational mission to their institution counted with much more significance. The study also discusses the implications relative to Herzberg's theory and offers elucidation as to what ultimately motivates academic librarians in their work, and reviews the findings of the research.

Susan, Baby and Sreerekha Pillai (2011) brought problems and prospects of professional development opportunities of academic library professionals in Universities in Kerala. The study is a part of the context of developments in information communication technology. The study recommends methods for improving the knowledge and skills of library professionals. The study evaluated the professional development activities of library professionals and their attitude towards continuing education programmes. Results of the analysis showed that the majority of the professionals have pursued higher degrees in library science or IT allied courses after entering the profession and that they have a positive attitude towards participation in training programmes and workshops and developments in ICT have a positive influence on the majority of library professionals' attitude towards continuing education programmes.

Ahmed et al. (2010) analyzed the effects of motivational factors on employees' job satisfaction from the non-academic staff of the University of the Punjab, Pakistan. The study intended to test Herzberg's motivation and hygiene theory in the background of non-academic staff and exploratory study based on collecting primary data. The study found that intrinsic motivational factors have a significant relationship with employee job satisfaction, whereas extrinsic factors have no significant relationship with employee job satisfaction. Besides, a significant difference was observed between gender, qualification, experience, job characteristics and job satisfaction.

Ghabra., Mansouri and Huda (2010) highlighted that motivation leads people to act or perform through a study of motivation techniques used by managers of Kuwait University libraries to improve employees' job satisfaction levels. The study examined techniques and issues that motivate library staff and how these differ among professionals and non-professionals. It appears that high productivity depends on the level of library workers.

Jana and Panigrahi (2010) found that college libraries suffer from a lot due to lack of proper motivation among librarians and identified causes of poor motivation towards work and suggests motivating parameters of job life for non-government college librarians in West Bengal. The study's finding is that prevalent factors like status, salary, service conditions, recruitment and promotional policy, and managerial relations are responsible for poor motivation.

Kacherki and Konnur (2010) highlighted that the term motivation is used to apply to the entire class of drives, desires, needs, wishes, and similar forces, which channelizes individual and group behaviour to achieve individual and organisational goals. The study examined some of the motivational factors for increasing work efficiency. An organisation's tasks motivate its employees to perform the work assigned to them so that the employees are satisfied and organizational goals are achieved. To make employees satisfied and committed to their job in academic libraries, there is a need for strong and effective motivation at the various levels and sections.

Aziagba (2009) surveyed library paraprofessional staff in Southern Nigeria on motivation for in-service training. The findings reveal that they are eager to take such opportunities, believing that in-service training will positively affect the prospect of their professional development. It also indicates that the paraprofessional staff believes that staff development or opportunity for further education is a highly motivating force that enhances productivity. Staff development is generally considered a positive source of staff motivation. The institutions also stand to gain as paraprofessional staff helps to provide essential library services.

Tella, Ayeni and Popoola (2007) investigated the perception of work motivation in relation to job satisfaction and organizational commitment of library personnel in academic and research libraries. The result indicates that both job satisfaction and commitment correlate with the perception of work by library workers. There is no significant difference exists in the perception of work motivation of professional and non-professional library personnel. Employees in academic libraries and research libraries have the same level of job satisfaction, and years of experience have no relationship with commitment.

Badu (2005) compared the staff's motivational preferences indicates that age groups and professional positions of the workers in a survey determine their motivational factor preferences. Respondents from the two sites have different motivational values. Therefore, various strategies are required to motivate the workers, and it is suggested that library managers avoid the assumption that what motivates them motivates the employees.

Toth (2005) surveyed the Hungarian Library Institute among LIS students at four LIS departments. The questions asked to LIS students are: what motivates students to do LIS training; which subjects and jobs are the students most interested in; after graduating, do the students want to work as librarians; and how the training may have influenced their original career plans. The results showed marked differences in responses between students of different years. However, it was clear that the training did positively influence students' commitment to the library profession.

Pors and Johannsen (2002) pointed that there is a growing societal concern about management and leadership qualities, and it is evident that leadership attributes are becoming increasingly important in the information sector. The survey focuses on various topics connected with leadership qualities and perceptions of different future challenges. The study focuses on job satisfaction among library directors. Job satisfaction is a central topic for motivational theories, and some of the more major of these are evaluated in relation to their findings. The analysis correlates job satisfaction with both extrinsic and intrinsic factors and the existence of major differences between these factors and the level of job satisfaction among library directors. Some of these factors appear to be connected with the level of activity in the library.

Williamson (1998) conducted a workshop that provides suggestions for motivating staff during the time of change. Major suggestions presented were giving staff autonomy, providing them with the training necessary to feel competent as their jobs change, and fostering feelings of being connected and the library. The workshop applies not just to libraries and serial units but to all other emerging areas of change. It describes that the key to a happy workplace is a happy staff, and managers must face change to accommodate effective management and earn respect and trust.

Abifarin (1997) investigated the level of motivation among librarians and paraprofessional staff. Major variables considered were: training opportunities, frequency of development, work environment, promotion prospects, sabbatical leave, communication and management style. It also attempts a brief definition of motivation and examines the views of a few classical writers on the subject. The study suggested some practical ways of motivating staff. Chief among them was the management style of university librarians going out to see what is going on in the various sections of the libraries. By this, university librarians will be able to observe their staff at work, talk to them in their various offices regularly, and carry out the spot assessment, listen, and praise staff.

Srivastava (1997) focused on Abraham Maslow's Need Hierarchy Theory and Fredrick Herzberg's two-factor theory. The study identified seventeen factors and prepared an itemised questionnaire, and analyzed. The responses did not reveal uniformity about the level of satisfaction or dissatisfaction.

Boers (1997) pointed out that formal mentorship programmes are instituted in libraries for three primary reasons: to bring new library school candidates into the profession, teach newly recruited librarians about their jobs, and help academic librarians prepare for the tenure review process. At the General Libraries at Emory University, Georgia, a different, inclusive programme was established, demonstrating the immense leadership potential of general staff and instituted not only to provide training and educational opportunities to the library staff but also to provide a communication bridge between general staff and librarians.

Rowley (1996) identified some issues that affect the motivation of staff in libraries and argues that such motivation is central to a quality culture. As libraries become more sophisticated in their approach to quality, self-motivation will become a central issue. Briefly reviews rational-economic model, social model, self-actualizing model, and complex model to review staff motivation. Environmental factors that impacted motivation included approaches to financial rewards, culture and the diversity of staff experience and roles. Possible strategies for motivation include developmental strategies such as appraisal, managing dissatisfiers and financial and social rewards.

Edem (1994) examined the quantity of publication output among librarians in Nigerian Universities. The study showed a low level of publication output among University librarians of Nigeria and identified barriers that inhibit publication output. Underfunding of education, high inflation rate, poor working and living conditions, lack of motivation and current information materials, and lack of time and interest for research are some of the factors that affect publication the output were studied. The study recommended motivational incentives and job satisfaction as a step towards performance effectiveness to enhance high publication productivity among librarians.

Antwi and Bello (1993) found that there are institutional and external factors that adversely affect the development of library professionals. Lack of promotion, absence of opportunity for training and education and unfavourable conditions of service are major negative factors, and promotional prospects in job would lead to increased productivity.

Thapisa (1993) reported that in motivating employees, first is to study the job content to enrich it. Job content includes intrinsic factors such as psychological growth, advancement, achievement, recognition, creativity and responsibility. It also includes skill variety, task identity, task significance, and autonomy. It needs to introduce fewer hierarchical structures and encourage more teamwork. It also needs to pay particular attention to payment systems. There should be clear knowledge about the comparative worth of some of the jobs offered in the library, and this we can do through a series of job analyses and evaluations. The pay that we offer in the end should not necessarily reflect the incumbent's position on the hierarchy but should consider other factors such as skill, responsibility, experience, effort, working conditions, risk and other hazards. This means that we need to develop wage and salary structures and policies through which we monitor attendance, productivity, and efficiency to provide proper guidance and direction.

Navalani (1990) gathered data on satisfaction with job characteristics of library and information from professionals and semi-professionals working in university libraries. Analysis of data revealed that most respondents are satisfied with most of the attributes of their work. But there are differences in the perception of men and women and seniors and juniors.

Lahiri (1988) conducted an empirical study based on Herzberg's theory of hygiene factors and motivational factors among library professionals relating to job satisfaction and motivational approach towards the system. The study outlined nearly negative in almost all areas of library services in the state. However, three platforms can play a significant role in improving the library network, and manpower has been slowly emerging and taking over leadership for the reorientation of the state.

Lynch and Verdin (1983) explored the relationship of sex, age, and tenure to the job satisfaction of library employees. The study results indicate that further library studies of job satisfaction will be more beneficial to the profession if placed within the context of the work environments in which librarians find themselves. The study demonstrates that job satisfaction can vary from one library unit to another and from one occupational group to another within the same library.

Katna and Saibaba (1982) attempted to recognize the role of motivation in libraries as in any other organization and its considerations in redesigning library job elements. The study analyses how library job elements may be motivating per se to workers. The concept of motivation to personnel should be applied in libraries to increase the organisation's efficiency.

Plate and Stone (1974) conducted a study based on Herzberg's theory. The findings from these experiments fall into the pattern provided by Herzberg theoretical framework. It indicates that the factors involved in producing job satisfaction and motivation are distinct and different from the factors that lead to job dissatisfaction. The factors creating job satisfaction and motivation are concerned primarily with the actual job content or work-process factors. The reasons for dissatisfaction or hygiene factors deal primarily with factors relating to the context in which the job is done in the job environment. Both sets of factors are closely interrelated. The study results suggested that these generalizations apply with as much force to librarianship as to other occupations on which Herzberg based his conclusions.

Conclusion

The review of the literature supports the research problem selected for the study. The purpose of the literature review is to review the development in the area of job motivation. It helps to point out those factors that potentially affect job motivation and its impact on various aspects in the library and information centres. The literature indicates that there has been a significant attempt in introducing job motivation in libraries. It enables to give a better understanding of the research problem selected. After reviewing the literature available, there have been very few studies on job motivation in the university libraries of Kerala. Therefore, the present study has been proposed as unique in this direction.

References

Abifarin, Abimbola. (1997). Motivating staff in Nigerian university libraries. Library Management, 18(3), 124 –128.

Ahmed, Ishfaq., Muhammad Musarrat Nawaz., Naveed Iqbal., Imran Ali., Zeeshan Shaukat., & Ahmad Usman. (2010). Effects of motivational factors on employees job satisfaction: A case study of University of the Punjab, Pakistan. International Journal of Business and Management, 5(3), 70-80. doi: http://dx.doi.org/10.5539/ijbm.v5n3p70

Alajmi,Bibi.,& Alasousi, Hessah. (2018). Understanding and motivating academic library employees: Theoretical implications. Library Management. https://doi.org/10.1108/LM-10-2017-0111

Ali Al-Aufi, Khulood Ahmed Al-Kalbani. (2014). Assessing work motivation for academic librarians in Oman. Library Management, 35(3), 199 – 212.https://doi.org/10.1108/LM-03-2013-0020

Alqudsi-ghabra., Taghreed, Mansouri., & Huda H. (2010). Staff motivation at Kuwait University Libraries. Journal of Information & Knowledge Management 9(4), 355-363.

Amune, Juliana Bosede. (2013). Job Motivation as a predictor of job satisfaction among professional and non-professional library staff in Ambrose Alli University, Ekpoma. International Journal Of Innovative Research & Development, 2(5), 1477-1497.

Antwi, I.K., & Bello, M.A. (1993). Motivation and productivity in academic libraries: A case study of the Abubakar Tafawa Balewa University Library. Bauchi Nigeria, Library Management, 14(5), 27-34.

Asante, Job; Yahaya, Wahabu & Yaahowaah, F. (2020). Motivating Employees using Equitable Pay Systems in Ghana: The Single Spine Salary Structure in Perspective. Asian Journal of Management.11(3),321-328.

Aziagba, Philip Chime. (2009). Staff development as motivation for paraprofessionals: A case study in southern Nigeria. Chinese Librarianship: An International Electronic Journal, (29), 1-8. Retrieved from http://www.iclc.us/cliej/cl27aziagba.pdf

Babalola, G. A. , & Nwalo, Kin. (2013). Influence of job motivation on the productivity of librarians in Colleges of Education in Nigeria. Information and Knowledge Management, 3(5), 70-75.

Badu, Edwin Ellis. (2005). Employee motivation in University Libraries in Ghana: A comparative analysis. Library and Information Sciences, 21(1), 38-46.

Bamgbose, Adeoye Augustine., & Ladipo, Sunday Olusola. (2017). Influence of motivation on academic library employees' performance and productivity in Lagos, Nigeria. Journal of Information and Knowledge Management., 8 (2), 33 – 47.

Baro, Emmanuel E. (2012). Job rotation program evaluation: The Niger Delta University Library. Aslib Proceedings, 64(4), 388 - 404. doi:http://dx.doi.org/10.1108/00012531211244743

Bernstein, A. M. (2011). A Herzbergian look at academic librarians and job satisfaction. Georgia Library Quarterly, 48(3), 6-21.

Boers, G. G. A. (1997). Mentorship and motivation: Librarians and general staff working together. Georgia Librarian, 34(4), 9-1.

Edem, Selong U. (1994). Quantitative measurement of publication output among librarians in Nigerian Universities. Annals of Library Science and Documentation, 41(2), 55-62.

Gairola, S. C. (2015). Assessment of Job Performance of Forest Guards in India: Relationship with Job Motivation, Satisfaction, Stress and other Critical Factors. Indian Forester, 141(3), 235-246.

Henczel, S., & Macauley, P. (2013, 20-23 October,). Professionalism: Exploring the role and responsibility of our professional associations. LIANZA Conference. Hamilton, New Zealand: Wai-Ora, Wai-Maori, Waikato. http://dx.doi.org/10.4236/oalib.1100804

Inceoglu, Ilke., Segers, Jesse., & Bartram, Dave. (2012). Age-related differences in work motivation. Journal of Occupational and Organizational Psychology, 85(2), 300–329. Retrieved from DOI: 10.1111/j.2044-8325.2011.02035.

Irfan,Mohammed P.P. , Haneefa,Mohammed, & Shyni, K.G. (2015). Workplace learning among library professionals of university libraries in Kerala. DESIDOC Journal of Library & Information Technology, 35(5), 376-381.

Jana, Lakshmikantha., & Panigrahi, Pijushkanti. (2010). Motivating parameters for betterment of library services: An evaluative study among non-government College Librarians in

West Bengal. Annals of Library and Information Studies, 57(1), 12-20.

Janelle R. Thomas., & Holley, R. P. (2012). Management versus repetitive tasks – avoiding "working for the weekend" A crash course in motivating library staff faced with seemingly endless tasks. New Library World, 113(9-10), 462 –473.

Kacherki, Umeshareddy., &Konnur, P.V. (2010). Motivational factors for increasing work efficiency among engineering College Librarians: A Comparative study of Maharashtra and Karnataka States. SRELS Journals of Information Management, 47(6), 681-688.

Katna, A.K., & Saibaba, B. (1982). Motivation of librarians. Library Herald, 23(2&3), 99-102.

Keogh, Patricia. (2013). Motivation for grant writing among academic librarians. New Library World, 114(3-4), 151 - 165. doi:http://dx.doi.org/10.1108/03074801311304069

Kont, Kate-Riin., & Jantson, Signe. (2014). A study of salary constraints in Estonian University Libraries. Slavic & East European Information Resources, 15(1-2), 60–79.

Lahiri, Ramansu. (1988). The human side of the library: A profile of Manipur. Annals of Library Science and Documentation, 35(2), 84-92.

Lamptey, R.B., Boateng, M.S., & Antwi, I.K (2013). Motivation and performance of librarians in public universities in Ghana. Library Philosophy and Practice (e-journal). Paper 911, 1-26. Retrieved from http://digitalcommons.unl.edu/libphilprac/911

Liz, Farler., & Broady-Preston, J. (2012). Workplace stress in libraries: A case study. Aslib Proceedings, 64(3), 225-240.doi.org/10.1108/00012531211244509.

Lynch, Beverly P., & Verdin, Jo Ann. (1983). Job satisfaction in libraries: Relationships of the work itself, age, sex, occupational group, tenure, supervisory level, career commitment, and library department. The Library Quarterly: Information, Community, Policy, 53(4), 434-447.

Madukoma, E., Akpa, V.O., & Okafor, U.N. (2014). Effect of training and motivation on job performance of library personnel of University of Lagos, Lagos State, Nigeria. Open Access Library Journal, 1,1-8. DOI: 10.4236/oalib.1100804

Mohammed, S., Tilahun, M., Kote, M., Mama, M., Tamiru, D., & Rosano, A. (2015). Validation of Health Extension Workers Job Motivation Scale in Gamo-Gofa Zone, Southern Ethiopia: A Cross-Sectional Study. International Scholarly Research Notices, 2015. https://doi.org/10.1155/2015/250610

Navalani. (1990). Job Characteristics of Library and Information work as satisfiers and dissatisfiers-A survey. Annals of Library Science and documentation, 37(3), 85-95.

Nazim, Mohammad. (2016). Knowledge sharing attitude of library professionals: A survey. Journal of Knowledge & Communication Management, 6(2),154-173. DOI: 10.5958/2277-7946.2016.00013.9.

Omol, Victoria Imabong (2020). Relationship among Job Satisfaction, Leadership Styles and Career Indecision of Library Personnel in Selected Private Universities in South-West, Nigeria. International Journal of Library Science, 9(2): 40-50.

Owolabi, K. A., Ajiboye, Bosede A., Bakare, Olubunmi.D., Bello, Taofeek.O., Omotoso, Akinkunmi. O., & Adeleke, Olateju. (2013). Librarians' attitude towards monetary and non-monetary incentives in university libraries: A case of selected university libraries in Nigeria. Annals of Library and Information Studies, 60(1), 22-26.

Plate, Kenneth H., & Stone, Elizabeth W. (1974). Factors affecting librarians' job satisfaction: A report of two studies. The Library Quarterly: Information, Community, Policy, 44(2), 97-110.

Popoola, S.O. (2007). Workplace, biographical and motivation factors affecting organizational commitment of records officers in Nigerian federal universities. African Journal of Library, Archives and Information Science, 17(1), 33-44.

Pors, Niels Ole., & Johannsen, Carl Gustav. (2002). Job satisfaction and motivational strategies among library directors. New Library World, 103(6), 199 – 209.

Raj, Sini S., & Devakumar, P.S. (2018). Motivation and job involvement among the employees of General Co-operative Marketing Societies in Kerala, International Journal of Management, IT & Engineering, 8(6), 253-269.

Rajpurohit, Rajpal Singh. (2016). Job satisfaction with regard to social recognition among the library professionals in Medical, Dental and Ayurvedic College of Rajasthan: A study. International Journal of Management, IT & Engineering, 6(12), 125-133.

Rowley, Jennifer. (1996). Motivation of staff in libraries. Library Management, 17(5), 31 – 35.

Sadeghi, S., & Sattari, S. (2018). A Comparative Study of Job Motivation and Work Engagement among Teachers Working in Smart and Ordinary Middle and High Schools in Parsabad. Journal of Ecophysiology and Occupational Health, 18(1-2), 12-17.

Sharma, Sanjeev., & Deolia, Ashish. (2017) Job involvement as a key component of work culture: A study based on survey of the library professionals of the special libraries of Delhi NCR. SSARSC International Journal of Library, Information Networks and Knowledge, 2 (1), 1-17.

Srivastava, Ashok. (1997). Work Motivation among school librarians in Jaipur: A comparative study of Government and Private Schools. ILA Bulletin, 33(1&2), 11-15.

Susan Mathew, K., Baby, M.D., & Sreerekha Pillai S. (2011). Professional development of academic library professionals in Kerala. Asia-Pacific Conference on Library & Information Education & Practice, 2011. 140-148.

Tella, Adeyinka., Ayeni, C O., & Popoola, S. O. (2007). Work motivation, job satisfaction, and organisational commitment of library personnel in academic and research libraries in Oyo State, Nigeria. Library Philosophy and Practice (e-journal), 1-16. Retrieved from http://digitalcommons.unl.edu/libphilprac/118

Thapisa, A.P.N. (1993). Structures, efficiency and motivation among Librarians. Library Management, 14(5), 20 – 26.

Toth, Mate. (2005). Career motivation of Hungarian library school students: a quick survey conducted among the full-time students of 4 LIS departments. Konyvtari Figyelo, 51(4), 705-733.

Williamson, Josephine. (1998). Happily ever after? Motivating staff in times of change. The Serials Librarian: From the Printed Page to the Digital Age, 34(3-4), 361-365.

Yahya, MSY., Wan Taib, W.M.A., Che Rusuli, M.S., Tasmin, R., & Hashim, N. (2012). Improvement of job design for Library Assistants. Proceedings of INTED 2012 Conference. 5-7 March 2012, (pp.2507-2514). Spain: Valencia.

3

Methodology and Profile of the Universities

Introduction

The present study was conducted among the regular Library and Information Science (LIS) professionals working in the eight-state universities in Kerala. Studies conducted on similar topics were collected after that the research design for the present study was developed. The researcher has visited all the selected universities in Kerala and met almost all the LIS professionals to collect the relevant primary data for the study.

The chapter elucidates research methods and techniques employed in the present study to answer the research problem. Research design, selection of variables, research methodology, and data analysis employed in the study are described in this chapter. The methodology adopted in the present investigation to answer the research problem designed for the study is included. The questionnaire method was used for the collection of data. The Census method of data collection was used for collecting data for the study. The study is organized around the following sections:

Research Design

The present study is aimed at investigating the role of some personal and motivational factors relating to the job in understanding the job motivation in the context of the libraries attached to the universities in Kerala. The study is conducted using a questionnaire for the collection of data. The study factors relating to universities and personal factors of the LIS professionals have been used as independent variables, and job motivation is used as the dependent variable.

(1) **Organizational and personal factors:** Organizational and personal factors include university, designation, education, sex, background, number of dependents, salary.

(2) **Job motivation factors:** Job motivation factors include factors related to six dimensions of job motivation. The factors are interest in work,

co-workers, promotion and growth, good pay, opportunities for higher studies and training, qualification and advancement.

Selection of Variables

Both dependent and independent variables on job motivation in the university libraries are identified and selected for conducting the study.

Research Steps and Methods

The research steps for investigating university libraries' job motivation in Kerala are described under the following headings.

Literature search

Formulating hypothesis

Organizational setup

Population of the study

Tools used for study

Data collection

Data analysis and interpretation

Testing of hypotheses

Generalization of hypotheses

Report writing

Literature Search

The basic methodology for the study adopted was to conduct a literature search in the area selected for the study. Literature search involves the use of primary, secondary and tertiary sources of information. They are books, periodicals, conference proceedings, reports, online sources such as EBSCO, ProQuest, Emerald, Google Scholar, J gate plus, j store, IFLA Conference proceedings, Shodganga, Science Direct, institutional repositories, online archives, books, conference proceedings, research reports, other web resources, DOAJ, JSTOR, DOAR, LISTA etc. are the relevant sources to the subject of the study.

Formulating Hypotheses

The study proceeds with the following hypotheses

H1: There is a lack of motivation among LIS professionals to carry out their duties and responsibilities.

H2: Authorities are adequately motivating the job of LIS professionals in the universities of Kerala.

H3: Role of library associations is minimum in motivating the job of qualified LIS Professionals.

Organizational set-up of the State Universities

The study was conducted to investigate the job motivation among the LIS professionals working in the university libraries in Kerala. There are 14 state universities and one Central University in Kerala, and the details relating to the universities are depicted in table No. 3.1

Table 3.1: State Universities in Kerala

Sl.No.	Name of University	Location	Type	Established
1.	University of Kerala	Thiruvananthapuram	Multidisciplinary	1937
2.	University of Calicut	Malappuram	Multidisciplinary	1968
3.	Cochin University of Science and Technology	Ernakulam	Multidisciplinary	1971
4.	Kerala Agricultural University	Thrissur	Agriculture, Engineering	1971
5.	Mahatma Gandhi University	Kottayam	Multidisciplinary	1983
6.	Sree Sankaracharya University of Sanskrit	Ernakulam	Sanskrit and Vedic studies	1993
7.	Kannur University	Kannur	Multidisciplinary	1996
8.	Kerala University of Health Sciences	Thrissur	Medical, Paramedical and Health Sciences	2010
9.	Kerala Veterinary and Animal Sciences University	Pookode	Veterinary and Animal Sciences	2010
10.	Kerala University of Fisheries and Ocean Studies	Kochi	Fisheries, Climate Science	2011
11.	Thunchath Ezhuthachan Malayalam University	Malappuram	Malayalam Language and Literature	2012
12.	A P J Abdul Kalam Kerala Technological University	Thiruvananthapuram	Engineering, Management, Technology	2014
13.	Sree Narayana Guru Open University	Kollam	Multidisciplinary	2020
14.	Kerala University of Digital Sciences, Innovation and Technology	Thiruvananthapuram	Technology	2020

Of the above fourteen universities mentioned in Table 3.1, eight state universities selected for the studies have full-fledged university libraries. In contrast, libraries of other state universities in Kerala are at the fledging stage with skeletal staff. In the present study, twelve state universities are established in different parts of the state. The following eight universities in Kerala: University of Kerala (UoK), University of Calicut (UoC), Cochin University of Science and Technology(CUSAT), Kerala Agricultural University (KAU), Mahatma Gandhi University, Kerala (MGUK), Kannur University(KU), Sree Sankaracharya University of Sanskrit (SSUS) and Kerala Veterinary and Animal Sciences University (KVASU) - have very strong University / Central Libraries, separate department libraries/constituent college libraries and study centre libraries. Therefore, the job motivation of the libraries of these universities is taken for the study. The libraries selected for the study are Kerala University Library (KUL), University of Calicut Library, CHMK Library (UoCL,), Mahatma Gandhi University Library (MGUL), Cochin University of Science and Technology Library (CUSATL), Kannur University Library, Dr. Hermann Gundert Central Library) (KUL) Sree Sankaracharya University of Sanskrit Library (SSUSL), Kerala Agricultural University Library and Information System (KAULIS) and Kerala Veterinary and Animal Science University Library (KVASUL). The newly established universities are exempted from the study. Moreover, the study is confined to the professional workforce in the central/university libraries of the universities and departmental libraries/constituent college libraries; all of them have qualified professional staff to manage.

Data used for the proposed objectives of the present study are collected from the libraries of the following eight state universities in Kerala.

1. University of Kerala, Thiruvananthapuram
2. University of Calicut, Thenjippalam
3. Cochin University of Science and Technology, Kochi
4. Mahatma Gandhi University, Kottayam
5. Kerala Agricultural University, Vellanikkara
6. Sree Sankaracharya University of Sanskrit, Kalady
7. Kannur University, Kannur
8. Kerala Veterinary and Animal Sciences University, Pookode

Population of the Study

The study population consists of a different cadre of LIS professionals working in the eight-state universities in Kerala. There is 385 sanctioned post of LIS professionals in the selected universities in Kerala. The Census method was used to collect data for the present study. The total population in the study, presently working in the universities, is taken together and consists of 276. The respondents under the study are only 266 professionally qualified and regular LIS professionals.

To ascertain the job motivation of LIS professionals in universities in Kerala, data were collected by the investigator personally, with the help of a questionnaire. A list of university libraries in Kerala was prepared, for which the University Handbook of Association of Indian Universities was taken as the basis. The recommended size of population procedure was adopted in this study after conducting a pilot study. Every possible effort was made to contact almost all the professionals within the library or outside the library. Investigator had made fruitful attempts in this regard. Anonymity of the respondents and confidentiality of the responses was promised. Questionnaires were distributed among the LIS professionals ranging from the top post of University Librarian in charge to the entry cadre level of Library Assistant/Professional Assistant Grade. II.

Respondents of the Study

The present study was determined from the response of a sample of 20 LIS from the pilot study. The population of the study is selected through the census method. From the sanctioned posts of 385 LIS professionals, 276 LIS professionals are in the selected eight universities in Kerala at the time of the study. Questionnaires were distributed to all the 276 LIS professionals, out of which 266 questionnaires were duly filled and returned with a response rate of 96.37 per cent. Accordingly, 266 qualified and permanent LIS professionals working in the libraries of eight state university libraries in Kerala are the respondents of the study.

Field Experiences

During the present study, the researcher visited all the libraries of the selected eight universities of Kerala. The author also witnessed the rise of the Library and Information profession for the last several decades. The researcher came across the job motivation of LIS professionals who are working in the said universities. The researcher also examined various circumstances faced by the LIS professionals working in the selected universities of Kerala during these visits. Significant experiences perceived during the visit of the university libraries attached to the universities in Kerala.

Data Collection

The population for the study was selected by census method procedure from the sanctioned post of 385 sanctioned LIS professionals. The cadre of LIS professionals is University librarian in charge, Deputy Librarians, Assistant Librarians, Reference Assistants (HG)/Reference Officers (HG), Reference Assistants/Reference officers/junior librarians, Technical Assistants/ Professional Assistants Grade I, Library Assistants/Professional Assistants Grade II. Both Primary and secondary data were collected for this study. Primary data were collected through questionnaires from LIS professionals. The questionnaire was designed and distributed to get responses from the target group of the research. Respondents were asked to provide opinions on the variables of this research. Secondary data of the study were collected from guide books, brochures, annual reports, journals, dissertations, theses, statistics, textbooks, documents from the Government of Kerala and Central Government and websites of the university libraries and universities under study were examined and analyzed. Tertiary sources of information were also used for the study. Sources of information used include electronic, print and online resources.

Tools Used for the Study

Questionnaire

A well-structured questionnaire was used for collecting primary data for the study. The questionnaire is a popular instrument for collecting the required data. The questionnaire was designed to suit the objectives of the study and pre-tested before the final use. The objectives of the study necessitated the need for a questionnaire for data collection. The major step involved in designing a questionnaire is to convert research objectives into information areas. A five-point Likert scale is used to measure the job motivation of LIS professionals in the selected universities.

Based on the objectives of the research, study in related areas and comprehensive literature review, a structured questionnaire was prepared for the collection of data. The questionnaire was prepared for collecting data from the LIS professionals. The investigator personally visited all the libraries under study, and questionnaires were distributed personally among almost all the professionals of the libraries under study. The questionnaire was used for the collection of data from the LIS professionals during the period in the universities covered under the study.

Structure of Questionnaire

The data gathered through a structured questionnaire from the qualified and permanent LIS professionals in the libraries of universities in Kerala. The questionnaire was distributed among all the LIS professionals working in the eight university libraries, department libraries and constituent college libraries and study centres libraries. The questionnaire has two sections.

Section 1 deals with general information of university libraries and LIS professionals.

Section 2 contains questions relating to job motivation in libraries. This section contains statements relating to interest in work, co-workers, promotion and growth, good pay, opportunities for higher studies and training, qualification and advancement, the role of library associations in motivating the job of LIS professionals, and authorities' role in motivating the job of LIS professionals.

The following methods were adopted to collect various types of data and information needed for the present study. It is presented in Table No.3.2.

Table 3.2: Methodology adopted for Collection of Data

Sl.No.	Scales	Authors	Total Respondents
1.	Job motivation	Agarwal (1987)	266

Job Motivation Scale

Job motivation refers to the degree to which a LIS professional has identified himself/herself with his/her work. This scale used for measuring involvement of a person in their job, and it was developed by Agarwal (1987). The scale comprises items each to be rated on 5-point scale ranging from strongly agree (1) to strongly disagree (5). The necessary modification was used to measure the variables job motivation for the purpose of the study. The scale consisted of relevant statements relating to the area concerned. The total score of each respondent was obtained by summing up the scores on all the items.

Procedure

The questionnaire was prepared to keep in view the insights and experiences gained during the pilot study run. Special attention was given to clarity of the meaning contained it each item. A total of 276 questionnaires were distributed to the respondents, and the 266 questionnaires were duly filled and returned.

Questions that usually are asked are including(a) how much time was required for the respondent, (b) how the researcher avoided introducing his own biased into the situation concerned (c) how the cooperative to the questions related to research was throughout, and (d) the impact of factors on the responses. All these questions are relevant because of the impact of one's understanding of the quality of the data and validity of questions.

Data Analysis and Interpretation

The data collected from the respondents were scored, tabulated and analyzed using appropriate statistical techniques.

Tabulation

In the study for tabulation, a statistical facility was used. The analysis and interpretations are classified, and tabulated data were made using SPSS Version 23 and Microsoft Excel. The data are entered in each column specified for each question in the questionnaire.

Statistical Tools used for Analysis

The statistical data package SPSS Version.23 was used, and the results obtained were interpreted statistically. Appropriate mathematical and statistical tools were used to analyze the data and draw relevant inferences. The analysis involves estimating the values of parameters collected from the population, testing the hypothesis for drawing inferences, examining the objectives, and finally reaching a conclusion. The data collected from the respondents of the study were scored, tabulated and analyzed using suitable statistical methods.

Keeping because of the study's amenability, hypotheses, and objectives, the data collected for the study were subjected to different statistical tools. The statistical tests included descriptive statistics, t-test, F test, and were used for the study. A brief description of the statistical tools used in the study is particularized below.

Descriptive Statistics

Descriptive statistics are brief descriptive coefficients that summarize a given data set, representing the entire or a sample of a population. Descriptive statistics are mean, median, standard deviation, minimum and maximum. Measures of central tendency include the mean, median, and mode, while measures of variability include the standard deviation, variance, minimum and maximum variables. The arithmetic mean scores for all variables were worked out to make suitable comparisons wherever necessary. Percentage analysis was done to classify the respondents selected into appropriate ground-based on the total score.

One-Sample t-Test

One sample *t*-test is a statistical procedure used to determine whether a process with a specific mean could have generated a sample of observations.

Descriptive statistics were used to identify the factors that lead to employee motivation. It was measured through a five-point scale. One sample t-test was used to compare the components with the mean.

Testing of Hypotheses

For testing the first hypothesis mean, median and standard deviation, the t-test is used. There is the above and average level of motivation among LIS professionals in the university libraries of Kerala.

Generalization of Hypotheses

Based on the testing of hypotheses, the researcher generalized the findings from job motivation in the university libraries in Kerala.

Report Writing

The study report or research report is new information or finding to the problems based on statistical analysis, and the researcher put forwarded suggestions.

Profile of the Universities

University of Kerala, Thiruvananthapuram

The University of Kerala was founded as the University of Travancore in the erstwhile princely State of Travancore in 1937. The university has sixteen faculties and forty-one departments of teaching and research and study centres and other departments. Teaching, Research and Knowledge extension are the mandates of the Departments. The university library as Travancore university library established in 1942. It was renamed Kerala University Library (KUL) in 1956. The Kerala University Library was started functioning in the present building in 1962.

The Kerala University Library (KUL), situated in the Thiruvananthapuram. It is the oldest university library in Kerala. It stocks over 3,50,000 books and subscribes to nearly 500 journals/ periodicals/ magazines. The library has a good collection of theses. It offers digital information services such as UGC-INFONET. Among its special collections, Kerala Studies is a unique one. The special collections also include Women's Studies, Government Publications, General Biographies, bound volumes of newspapers and journals and rare books. It is the only library in Kerala that serves as a United Nations (UN) and World Bank publication.

The Kerala University Library is complemented with several specialised Department Libraries and Campus Library at Karyavattom and Study Centre Libraries at Alappuzha, Kollam and Pandalam. The University Institutes of Technology (UITs), University Teacher Education Centres (UTECs), and University College of Engineering (UCE) also have specialized libraries. Library membership is open to faculty, students, research scholars, non-teaching staff of university, graduate public, very important officials and institutions.

Mahatma Gandhi University, Kerala, Priya Darsini Hills, Kottayam

The Mahatma Gandhi University, Kerala was established in 1983. The University conducts programmes at the Undergraduate, Postgraduate, M.Phil. and Doctoral levels through its 17 University Departments, 1 International and Inter University Centre, 7 Inter-University Centres, 10 Inter School Centres, 77 Government/Aided Affiliated Colleges including 10 Autonomous Colleges (of which 8 are Colleges with Potential for Excellence), 200 Unaided Affiliated Colleges and 273 Recognized Research Centres.

The University library was started in 1989. The library, from its infancy, has taken exclusive initiatives to imbibe ICT into the entire gamut of its activities. MGU Library and Information System consist of University Central Library, more libraries of the departments/schools and 4 Study Centres situated on different campuses. The university central library is situated on the main campus of the university at Athirampuzha, Kottayam. University library is effectively catering to the academic needs of faculty members, students and research scholars of departments/schools of teaching and research and affiliated colleges.

The library is a member of the INFLIBNET Centre, Ahmedabad. Through UGC-INFONET, E-journals Consortium, the Library has access to e-journals and databases. The Library also has access to major online databases. It adopts Open Access of its doctoral research theses to academics worldwide. The University Library has a collection of 57,350 books, 2,500 e-books, 2,050 theses and 7,500 bound volumes of journals. In addition to this, the library subscribes current Indian and Foreign journals, Online Databases and Online Journal Archives.

Cochin University of Science and Technology (CUSAT), Kochi

The University of Cochin established in 1971. The University of Cochin was reorganised and converted into a full-fledged University of Science and Technology by the Cochin University of Science and Technology Act, 1986 to promote graduate and post-graduate studies and advanced research in Applied Sciences, Technology, Industry, Commerce, Management and Social Science.

The Cochin University Library was established in 1977 to meet the information requirements of the university's academic community and for the collection, organization, and dissemination of information. The library offers information based on the printed sources and e- sources. It caters to the academic needs of faculty members, students and research scholars of departments/schools of teaching and research. The library has built up a collection on all branches of Science and Technology.

The library has been exploring all the possibilities and advancements of technological applications. The library became automated. The resources of the library are open to all the departments. The Library collection includes several reference books, doctoral theses and back volumes of periodicals, patents, etc. The library offers loan service, inter-library loan, literature search, compilation of bibliographies, reprographic service, conference alert service, new addition alert service and consultancy service.

The library has IP enabled access to electronic databases, e-journals and online books, which in turn assist the productivity of the university academics. The library provides access to periodicals, online journals and databases, e-books. Online access is also provided to citation and scientometric databases. The library also facilitates access to several electronic journals. Members can avail of the service of DELNET, UGC-INFONET, JCCC portal. The bibliographic records of documents can be accessed through the web OPAC facility. The library has the union catalogue of all the libraries of CUSAT.

The special collection of the university includes the collection of theses, patents, art books, WHO and other UN documents, Census reports, Vikasana Rekha (Kerala Development Report) and books on Library and Information Science.

The library houses an Institutional Repository, "Dyuthi". A research gateway to publications of CUSAT faculty is provided through YUJ, a bibliographic information system. A media repository of various events, lectures etc. are at the University.

University of Calicut, Tenjippalam, Malappuram

The University of Calicut, the largest university in Kerala, was established in the year 1968. The University aims to nurture excellence in education and research in its catchment areas of Northern Kerala. The university emphasises fostering quality human resources and promoting productive research that benefits local communities and wider humanity.

The University of Calicut Library (UoCL) started functioning in 1971. The present library building started functioning in 1974. In 1993 the name of the university library was renamed after C. H. Mohammed Koya (the former Chief Minister and Minister for Education, Government of Kerala) as CH Mohammed Koya Library. It is primarily concerned with the conservation and dissemination of knowledge to its users. It plays an important role in the provision of information to the academic community in the Malabar region. A part from the university library, a study centre library functions in Kozhikode city to serve the academic community in the Kozhikode city and around. Library membership is free to the university teachers, students, research scholars, non-teaching staff of the university, and some categories. The library has a rich collection of books, journals, bound volume of journals, maps, microfiche, CDs, theses, dissertations etc. In addition to the UGC-INFONET e-journal consortia, library has electronic journals and e-books.

Sree Sankaracharya University of Sanskrit, Kalady, Ernakulam

The Sree Sankaracharya University of Sanskrit (SSUS) was established in 1994 in Kalady, Kerala. It is one of the pioneering Sanskrit Universities in India. The foundational aim of the University is to impart knowledge in Sanskrit, other Indian and Foreign Languages, Social Sciences and Fine Arts. As an institution imbibing multiple cultural heritages and housing both traditional and modern knowledge systems and art forms, the university envisions a harmonious advance in society through educational excellence. The Sree Sankaracharya University of Sanskrit has eight regional campuses across Kerala.

Sree Sankaracharya University of Sanskrit Central Library has been functioning in the main campus at Kalady since the university's inception in 1993. The Central Library aims to develop a comprehensive collection of documents for the faculty, students and research scholars in their educational and research activities. The library has a collection of ancient and modern thoughts, mainly on Social Science and Humanities. The university's collection includes books, Manuscripts, Journals and Periodicals, CD ROMs, Thesis and Dissertations.

Sections of the university library include Circulation, Acquisition, Technical, Reference, Journal, Magazine, Newspaper, Maintenance, Information Technology section. Books are arranged on the shelves according to the Dewey Decimal Classification (DDC) Scheme. Online Public Access Catalogue (OPAC) is available in the library. The services in the library include display of new arrivals, reservation, renewals, photocopying and CD-ROM search. In addition to the university library, the university has regional campus libraries at Thiruvanathapuram, Panmana, Thuravoor, Ettumanoor, Thrissur, Tirur,

Koylandy and Payyanur. Libraries of the university are adopting the latest technology. It facilitates quick access information resources. UGC-INFONET project provides access to electronic journals and database. The library has a number of rare and historically important manuscripts and palm leaves in different languages.

Kerala Agricultural University, Vellanikkara, Thrissur

The Kerala Agricultural University (KAU), established on 24th February 1971, started functioning on 1st February 1972. Kerala Agricultural University, at present, has eight constituent colleges. The colleges are College of Agriculture, Padannakkad, College of Agriculture, Vellanikkara. College of Agriculture, Vellayani, College of Agriculture, Wayanad, College of Climate Change and Environmental Science, Vellanikkara, College of Co-operation, Banking & Management, Vellanikkara, College of Forestry, Vellanikkara, Kelappaji College of Agricultural Engineering & Technology, Tavanur, Besides, Institute of Agriculture Technology RARS, is situated at Regional Agricultural Research Station, Pattambi. All of them are having good library facilities.

Now the Kerala Agricultural University has six Regional Agricultural Research Stations, seven Krishi Vigyan Kendras, 15 Research Stations and 16 Research and Extension Units under the faculties of Agriculture, Agricultural Engineering and Forestry are also functioning in Kerala Agricultural University. The university has been discharging the mandatory functions of initiating, interfacing and integrating Education, Research and Extension activities in agriculture and allied branches of science.

The unit for Library and Information Services was established in 1995 at the Kerala Agricultural University. The Central Library is located at the main campus Vellanikkara. It came up in 1995. Kerala Agricultural University Library and Information System (KAULIS) was inaugurated in 1998. Kerala Agricultural University Library and Information System (KAULIS) is envisaged to collect, organise, and disseminate information on agriculture. KAULIS consists of the University Central Library, libraries in all Constituent Colleges and Research Stations.

The Central Library is equipped with state-of-the-art hardware, software, power backup and IT Labs for modern library and information services and WiFi facility. Information resources in print and electronic media cover all subjects in the thrust areas of the university plus subjects of general interest. Central Library is the repository of M.Sc. / Ph.D. theses submitted to various faculties of the University. The library has a collection of back volumes of Indian and Foreign Journals. The library has accessibility to Consortium for e-Resources in Agriculture (CeRA), India Agristat, Centre for Agriculture and Bioscience (CAB) International, CABI e-books, Science Direct e-books, Indian e-books, DELNET, AgriCat, catalogue and Krishikosh, a digital repository.

The library services include Book Loan Service, Reference Service, Literature search, Reprographic Service, Current Awareness Service and Selective Dissemination of Information, Document Delivery Service, Database and Internet-based Services and Inter Library Loan Service.

Kannur University, Mangalathuparamba, Kannur

Kannur University was established in the year 1996. The University by the name "Malabar University" had come into existence on 9th November 1995. The Kannur University was started with the objective to establish a teaching, residential and affiliating University, promoting the development of higher education in Kasargod and Kannur revenue Districts and the Mananthavady Taluk of Wayanad District in the State of Kerala. Kannur University is a multi-campus university with campuses at Kannur, Kasaragod, Mananthavady, Payyannur, Thalassery, Kanhangad and other places necessary for providing study and research facilities to promote advanced knowledge in Science and Technology and other relevant disciplines.

Kannur University Central Library started functioning at Kannur in 1998. It was opened to the academic community in 1999. It was renamed Dr. Hermann Gundart Central Library in 2016. The University Central Library plays a pivotal role in supplementing and complementing the teaching and learning process and research activities.

Mission of the library

To acquire, organize and provide access to various kinds of information sources, including books, journals, theses etc., both in print and electronic forms and provide excellent IT-based information services to the student, researchers and faculty to pursue academic excellence in their field.

Goals of the Library

To provide an information-rich environment that supports and encourages excellence in research and scholarship.

To provide value-added services to the library members.

To meet the customer needs in time.

To ensure continuous improvement in library services and products.

To offer effective, responsive, and friendly staff-user interface.

The library has been serving the various sections of the University community viz., students, research scholars, faculty members, staff, graduates and students belonging to other Universities in the jurisdiction of Kannur University. The Govt. of Kerala has sanctioned Rs. 25 crores to the Kannur University Central Library and Research Centre. Neo literate section, the section for visually challenged, audiovisual section, archives section for Malabar collection, art gallery, heritage museum, digital library will also be provided in the library.

The library provides the services of DELNET and UGC-INFONET e-journal consortium. It has a collection of books, theses, dissertations, databases, consortia-based resources, CD-ROMs, e-journals, e-books, back volume of journals, currently subscribing journals. The library has a special collection of Thaliyola. In addition to the Central Library, campus libraries function in the different campuses at Kasaragod, Nileshwaram, Payyannur, Mangattuparamba, Thalassery and Mananthavady.

Kerala Veterinary and Animal Sciences University, Pookode, Wayanad

The Government of Kerala established the Kerala Veterinary and Animal Sciences University (KVASU) in December 2010 to further education, research and extension services in veterinary and Animal Sciences with headquarters at Pookode. The University encompasses three faculties- the Faculty of Veterinary and Animal Sciences, the Faculty of Dairy Science and the Faculty of Poultry Science and eight constituent colleges. The University has a wide network of instructional and research farms spread throughout the state.

In KVASUL, there is no central library but various libraries attached to the university's constituent colleges. The library's primary mission is to support the educational and research programs of the Kerala Veterinary and Animal Sciences University by providing access to information consistent with the present and the anticipated educational and research functions of the institute. The secondary mission is to serve as a resource centre for the scholars and scientific community of the Veterinary and Animal Husbandry arena.

The web enabled University Library has a vast collection of books, journals, and online access to e-journals. Open-source software is using in institutional repository and library management. This library offers reference and referral services, Circulation Service including Inter-Library Loan and Document Delivery Service, CAS & SDI, display of recent additions of periodicals and books, Literature search, Internet Browsing, Consultancy Service, User guidance & education etc.

College of Veterinary and Animal Sciences, Mannuthy was established in 1955. It is one of the constituent colleges of the Kerala Veterinary and Animal Sciences University. The Library of the College of Veterinary and Animal Sciences, Mannuthy, serves students, faculties and Research Fellows to refer to various materials related to Veterinary and Animal Sciences. The library has books, journals, electronic materials, Theses, Magazines and other reference materials. The library also has a facility where the members can get the latest news and updates about the respective subject.

Conclusion

The chapter has presented details of the research methodology and profile of the universities selected for the study. The variables used for the study, tools and procedures for data collection and the statistical techniques used for data analysis were also particularized. The data collected from the LIS professionals working in the different universities were collected and analyzed using appropriate description, inference, and conclusion. The detailed view of the analysis and its interpretation by applying methodology and statistical tools is presented in chapter V.

References

Agarwal, U. N. (1987). Measuring job-involvement in India. Indian Journal of Industrial Relations, 14, 219-31.

American Psychological Association. (2010). Publication manual of the American Psychological Association (6th ed.). Washington, DC: Author.

hgcl.kannuruniversity.ac.in

http://library.kau.in/in

https://library.ssus.ac.in

https://www.kvasu.ac.in/

KAU Handbook 2021. (2021). Thrissur: Kerala Agricultural University

library.cusat.ac.in

library. mgu.ac.in

Parasuraman, A, et al. (1988). SERVQUAL: A multiple item scale for measuring consumer perceptions of service quality. Journal of Retailing. 64(1), 12-40.

Payne, R. L., & Pheysey, D.C. (1971). G.G. Stern's organizational climate index: A reconceptualization and application to business organizations. Organizational Behavior and Human Performance, 6, 77–78.

SPSS. (2016). Wikipedia. Retrieved from http://en.wikipedia.org/wiki/SPSS.

Walton, R. (1975). Criteria for quality of work life. In Davis, L.E and Cherns, R.L (Eds.) Quality of work life: Problems, prospects, and the State of the Art (Vol. I, pp.12-15). New York: Free Press.

www.kulib.in

4

Job Motivation in Libraries An Overview

Introduction

Human behaviour is a complex phenomenon. Several variables influence it. The relationship between man and workplace is of paramount importance. Several factors motivate the working environment of the personnel in an organization. Research is going on the motive of people at work. Productivity of Library and Information Science (LIS) Professional depends on several factors. Job motivation affects their productivity in libraries. The present chapter tries to examine the factors that necessitate job motivation. The university libraries are regarded as service institutions. University libraries are changing their services for the benefit of the user fraternity.

Job Motivation

Motivation is a psychological process. Job motivation refers to how urges, drives, desires, aspirations, strivings, or needs direct, control, or elaborate the behaviour of human beings. Motivation is the complex force starting and keeping a person at work in an organization. Motivation moves the person to actions and continues him in the course of action already initiated. Libraries need a stable workforce of talented and motivated employees. Personnel committed to their organizations are more likely to put extra effort, accept change even in adversities for achieving high-quality work. Job motivation converts the ability to work into a will to work. It raises the level of performance of employees in the organization.

Motivation has an important role in human life. It has it impact on the work situation also. Performance is proportionally related to motivation. High motivation leads to high performance and low motivation to low performance. It is pertinent to mention that "you can lead a horse to water, but you can't make it drink" (proverb). It reveals that you can give someone an advantage or provide them with an opportunity, but you can't force them to do something if they don't want to. It reveals the importance of motivation for any desired activity. People will not work unless they are prepared to, and the level of performance is most likely to be determined by the strength of motivation.

Definitions

The word "Job" means a piece of work to be done. It refers to a particular and defined task. Work motivation is a set of energetic forces that originate both within as well as beyond as individual's being, to initiate work-related behaviour, and to determine its from, direction, intensity, and duration (Pinder, 2008). A definition of motivation is that which makes people put real effort and energy into what they do. Psychologists have been exploring how to motivate employees since early in the last century, and a lot of knowledge on human motivation has been developed and widely applied (Srivastava and Barmola, 2011).

Employee motivation is a long-lasting research subject on Human Resource Management (HRM) and organizational psychology's remits of investigation. Employee motivation refers to the desire to adopt high levels of personal effort justified by the achievement of organizational objectives, leading these efforts to the satisfaction of a particular individual need (Gomes, Asseiro, and Ribeiro, 2013).

Types of Motivation

There are two types of motivation. They are intrinsic motivation and extrinsic motivation.

- **Intrinsic motivation:** Intrinsic motivation emerges within the person. Intrinsic motivation leads to a particular behaviour. The person sees it as aligned to his value systems and belief systems, which fulfil a desire that holds specific importance. Intrinsic motivation is in the relationship between perceived training opportunities and organizational citizenship behaviours. Miao and Evans (2007) examined the impact of motivation on role perceptions and job performance.

- **Extrinsic motivation:** Extrinsic motivation emerges when persons do something because of their external forces. This kind of motivation deals with the motivation which is generated by external desires. To further this, an incentive is the anticipated reward or aversive event available in the environment (Deckers, 2010).

Objectives of Job Motivation

Motivation is a person's internal disposition to be concerned with and approach positive incentives and avoid negative incentives. Motivation has continuously been an area of interest to managers of different organizations, including profit and service-based organizations. Most libraries have a mission to offer a high-quality service to all of their clientele. The purpose of motivation is to create a condition in which people are willing to work with zeal, initiative, interest, and enthusiasm, with a high personal and group moral satisfaction with a sense of responsibility. Motivation techniques utilized to stimulate employee growth. Motivation is critical in the establishment and further development of quality service. Motivation is a key element to librarians as to any other profession.

The objectives of job motivation are:

- **Increase willingness to do the job:** Increasing worker motivation means workers willingly carry out the tasks required by the job. This also means managers are less likely to pull teeth to get workers to perform tasks. The job description should match the duties, qualities and competencies of the position. This helps administrators and workers understand the position's demands before, during, and after coming into a position.
- **Increase performance:** Increasing worker motivation serves to help organizations meet internal and external goals. Increasing worker motivation has shown to increase job performance levels. Administrators should help workers feel as a part of a team in the organization. Increased worker performance generally means increased satisfaction of the clientele.
- **Increase retention:** An administrative team that includes the workforce in decision-making processes for strategic planning in achieving organizational goals has increased worker retention. Helping workers feel engaged throughout the process increases worker motivation and to feel invested in the organization. Growing worker motivation and job satisfaction have been linked to higher rates of retention and increased worker productivity.
- **Increase job satisfaction:** A poor working environment can make any good worker lose motivation. Job incentives such as competitive pay, benefits and programs such as tuition reimbursement, flexible working hours, scheduling and day-off show increased levels of job satisfaction among the staff. Since most workers do not stay in the same position for an entire career, training programs have increased job satisfaction. Increased job satisfaction has increased worker motivation, which then increases job retention rates, performance and productivity.

Scope of Job Motivation

It is important for organizations to understand and to structure the work environment to encourage productive behaviours and discourage those that are unproductive given work motivation's role in influencing workplace behavior and performance. (Jex and Britt, 2008).

There is a consensus that motivation involves three psychological processes: arousal, direction, and intensity. Arousal is what initiates action. It is fuelled by a person's need or desire for something missing from their lives at a given moment, either totally or partially. Direction refers to the path employees take in accomplishing the goals they set for themselves. Finally, intensity is the vigour and amount of energy employees put into this goal-directed work performance. The level of intensity is based on the importance and difficulty of the goal. These psychological processes result in four outcomes. First, motivation serves to direct attention, focusing on particular issues, people, tasks, etc. It also serves to stimulate an employee to put forth the effort. Next, motivation results in persistence, preventing one from deviating from the goal-seeking behavior. Finally, motivation results in task strategies. Patterns of behavior produced to reach a particular goal (Mitchell and Daniels, 2003).

Benefits of Job Motivation

Motivation plays a critical role through which employees feel a sense of commitment to the goals of organization. The extent of employees committed to the organizations will have an impact on the effectiveness of organizations. The satisfaction and dissatisfaction among the personnel working in the organizations are, both directly and indirectly related to motivation, job satisfaction and perception of organizational climate.

Theories of Job Motivation

Motivation theories can be classified on several bases. The major classification of job motivation theories cover

- **Natural vs Rational:** It is based on whether the underlying theory of human cognition is based on natural forces like drives, needs, desires etc. or some kind of rationality like instrumentality, meaningfulness, self-identity etc.
- **Content vs Process:** It is based on whether the focus is on the content "what" motivates vs process "how" motivation occurs.
- **Reinforcement theories:** Reinforcement theories focus on the ways in which behaviour is learnt.

Content Theories of Work Motivation

The content approach to work motivation is associated with Maslow, McGregor, Herzberg, Atkinson, and McClelland. These are familiar to managers because they have strongly influenced the management field and have affected the thoughts and actions of practising managers. The content theories of work motivation attempt to determine what it is that motivates people at work. The content theories of work motivation are concerned with identifying the needs drives that people have and how these needs drives are prioritized. They are concerned with the types of incentives or goals people strive to attain to be satisfied and performed. The content theories are static because they incorporate only one or few points in time or past or present-time oriented. Therefore, they do not necessarily predict work motivation or behaviour but are still important to understand what motivates people at work. At first, money was the only incentive in scientific management. Later, that incentives include working in conditions, security, and perhaps a domestic style of supervision of human relations. Recently, the content of motivation has been deemed to be the so-called higher-level needs or motivates, such as esteem and self-actualization (Maslow); responsibility, recognition, responsibility, achievement, and advancement (Herzberg) and growth and personal development (Alderfer). A thorough content of the major studies contributes to understanding and leads to some of the application techniques of motivation.

Process Theories of Work Motivation

The process approach emphasizes how and by what goals individuals are motivated. In this view, needs are just one element in the process by which individuals decide how to behave. Basic to process theories of motivation is the notion of expectancy. That is, what a person anticipates is likely to occur as a result of their behaviour. An additional factor in motivation is the valence or strength of an individual's preference for the expected outcome. The content models attempt to identify what motivates people at work, for example, self-actualization, responsibility and growth. They try to specify correlates of motivated behaviour. The process theories are more concerned with the cognitive antecedents that go into motivation or effort and, more important, with the way they relate to one another.

The process theories provide a much theoretical explanation of work motivation. The expectancy model of Vroom and the extensions and refinements how they relate to one another in the complex process of work motivation. The Porter-Lawler model gives specific attention to important relationship between performance and satisfaction. Porter and Lawler propose that performance leads to satisfaction instead of the human relation assumption of the reverse.

Conceptual and methodological problems remain. Unlike content models, expectancy models are relatively complex and difficult to translate into actual practice. Consequently, they have generally failed to meet goals of prediction and control of organizational behaviour.

Reinforcement Theories of Work Motivation

Reinforcement theories, associated the B.F. Skinner and others are also often called behaviour modification or apparent conditioning. These theories do not utilize the concept of a motive or a process of motivation. However, they deal with how the consequences of a past action influence future actions in a cyclical learning process. In this view, people behave the way they do because, in the past, they learnt that certain behaviours were associated with pleasant outcomes and that behaviours were associated with unpleasant outcomes. Because people generally prefer pleasant outcomes, they are likely to avoid behaviours with unpleasant consequences.

Contemporary Theories of Work Motivation

Work-motivation theories are generally categorized into content and process approaches. Equity and attribution theories have emerged in recent years and command the attention of most of the research. An understanding of these two theoretical developments is now necessary to study work motivation in organizational behaviour. Recently, equity theory and, especially, attribution theory have received increased attention in academic circles. Both theories-equity theory, which is based upon perceived input-outcome ratios, and attribution theory, which ascribes internal, external, and other causes to behaviour-lend increased understanding to the complex cognitive process of work motivation but have the same limitation as the expectancy models for prediction and control in the practice of human resources management. Attribution theory can make a potentially important contribution to the cognitive development of work-motivations theory. It may overcome some of the applicable limitations of the expectancy and equity theories of work motivation and bring us closer to the goals of prediction and control. Finally, from other disciplines, control and agency theories are mentioned as representative of other approaches receiving attention in organizational behaviour.

Abraham Maslow, Herzberg, Douglas McGregor, Victor Vroom, McClelland and a host of others have contributed to the theory of motivation. Satisfying factors motivate and dissatisfying factors deterrents. Motivating factors are achievement, recognition, the job conducted, responsibility, promotion, and the factors itself for professional development (Kishore, 1981).

The five motivational theories formulated in the 1950s by five eminent personalities have received considerable attention and frequently offered explanations of or approaches. The theories are given below:

1. Maslow's Theory of Hierarchy of Needs (1954)

Maslow's Hierarchy of Needs is a theory in psychology proposed by Abraham Maslow in his paper "A Theory of Human Motivation" (Maslow, 1943)). Maslow's theory was fully expressed in his book 'Motivation and Personality' (Maslow, 1954). The best-known theory is the Hierarchy of Needs proposed by Abraham Maslow. Maslow connects the creation of the existence of people's sense of satisfaction with the maintenance of the classified needs. These are psychological needs like eating, drinking, testing, etc., securities need like pension, health insurance, etc., the need to self-esteem like self-confidence, recognition, adoration, to be given importance, status, etc. need of self-actualization like maximization of the latent power and capacity, development of abilities, etc.

2. Herzberg's Motivation-Hygiene Theory (1959)

Herzberg's theory of work motivation is also called the two-factor theory of motivation. The framework of human behaviour studied by Herzberg 'Motivation-Hygiene theory or two-factor theory'. His hygiene factors cover management policy and administration, supervision, relationship with supervision, colleagues, subordinates, salaries, job security, working conditions and status. The factors are similar to Maslow's "lower-order needs", namely, physiological, security, and social needs. Motivational factors related to the job itself include achievement, recognition, advancement, work itself, the possibility of growth and responsibility (Maslow's higher-order needs viz. ego and self-actualization). These factors produce motivation and high job satisfaction. The present study attempts to view a cross-section of practising library professionals in the above framework and an attempt to know their present constraints and explore the possibility of getting rid of them.

3. Vroom's Expectancy Theory (1964)

The expectancy theory of motivation was established by Victor Vroom of Yale school of management in 1964 to believe that motivation is based on the expectation of desired outcomes. The theory is based on three concepts. The concepts are valence, expectancy, and force. Valence is the attractiveness of potential rewards, outcomes, or incentives. Expectancy is a person's belief that they will or will not be able to reach the desired outcome. Force is a person's motivation to perform, (Arlt and Toupence, 2007).

Expectancy theory explains the behaviour process in which an individual selects a behaviour option over another and why or how it is made related to their goal.

There is an equation for Expectancy theory that goes as follows:

M = E x I x V

OR

Motivation = Expectancy x

Instrumentality x Valence

M (Motivation) is the amount an individual will be motivated by the condition or environment they placed themselves in. It is based on the following hence the equation.

E (Expectancy) is the person's perception that effort will result in performance. In other words, it's the person assessment of how well and what kind of effort will relate to better performance.

I (Instrumentality) is a person's perception that performance will be rewarded or punished.

V (Valence) is the perceived amount of reward or punishment that will result from performance.

The theory depicts that the intensity of tendency to perform in a particular manner is dependent on the intensity of an expectation that performance will be followed by a definite outcome and on appeal of an outcome to an individual.

4. McGregor's Theory X and Y (1966)

Theory X and Theory Y are theories of human motivation and management. They were created and developed by Douglas McGregor at the MIT Sloan School of Management in the 1960s. Theory X and Theory Y describe two contrasting models of workforce motivation applied by managers in human resource management, organizational behaviour, organizational communication and organizational development. As per these models, the two opposing sets of general assumptions of how workers are motivated to form the basis for two different managerial styles. Theory X stresses the importance of strict supervision, external rewards, and penalties. Theory Y highlights the motivating role of job satisfaction and encourages workers to approach tasks without direct supervision.

McGregor believed that Theory Y assumptions best captured the true nature of workers. He argued that authority should free up their employees to unleash their full creative and productive potential. McGregor theory of Y assumptions applies to the environment of the library for better job satisfaction. Theory Y advocates capturing the best true nature of workers and should guide management practice. It is argued that the authority concerned should give the library professional liberty to develop their potentiality fully.

5. McClelland's Needs Theory (1975)

McClelland is credited with developing the Achievement Motivation Theory. It is commonly referred to as need achievement or n-achievement theory (Hoy and Miskel, 2008). Need theory, also known as Three Needs Theory proposed by psychologist David McClelland. It is a motivational model that attempts to explain how the needs for achievement, power, and affiliation affect people's actions from a managerial context. This model was developed in the 1960s, soon after Maslow's Hierarchy of needs. McClelland stated that all have these three types of motivation regardless of age, sex, race, or culture. The type of motivation by which each individual is driven derives from their life experiences and the opinions of their culture.

David McClelland has proposed that there are three major relevant motives or needs in workplace situations:

- The need for achievement: It is the drive to excel, achieve in relation to a set of standards, and strive to succeed.
- The need for affiliation: It is the desire for friendly and close interpersonal relationships.
- The need for power: It is the need to make others behave in a way that they would not have behaved otherwise.

McClelland's three needs are non-sequential but instead are used in relation to each other.

All these theories are related to library and information science because these different theories highlight the importance of job satisfaction and motivation among library professionals.

Managers of libraries and information centres can increase the expectancy level of employees by adopting several steps. The usual methods adopted for increasing the expectancy level are as follows.

- Increasing level of confidence of employees to achieve the required level of performance

- Praise, either public or private
- Rewards for sincerity, dedication and constructive ideas
- Participation in decision making
- Delegation of responsibility
- Increase in the pay
- Improved work environment
- Job enrichment
- Job enlargement
- Good Organizational Climate
- Job motivation, and so on

Factors of Job Motivation

Several studies relating to job motivation pointed out that motivated workers are more productive. Both financial and non-financial factors influence motivation. The main motivating factors of the library staff were appreciation by authorities, colleagues, clientele and the community, a stable job and income and training. The main discouraging factors were related to low salaries and difficult working conditions.

The dimensions of the job identified by different researchers are as follows:

- **Work:** Work is an activity that requires a persons' mental or physical effort. Motivating to work is a process to energize employees to the work goal through a specific path. It deals with how the job provided the individual with tasks, opportunities for learning and the chance to accept responsibility.
- **Pay:** Pay level is regarded as an important indicator for status, success and happiness. It deals with the amount of financial benefit that one receives and the degree to which this is viewed as equitable to others in the organizations.
- **Promotion:** Promotion rejuvenates an employee with more enthusiasm and responsibility. It helps to perform effectively. It deals with the chance for advancement in the organization.
- **Supervision:** Supervision is an act or instance of directing, managing or oversight. It deals with the activities of the higher authority to provide technical assistance and behavioural support.

- **Co-workers:** Co-workers encourage a shared working environment. It deals with the degree to which the fellow workers are technically proficient and socially supportive. One can motivate one's co-workers to perform better.
- **Purpose:** The purpose is a commitment to carry out an action in the present or actions in the future. The supervisor should undertake to align the individual's purpose with that of the organization.
- **Expectation:** The expectation is a belief about what might happen in future. Communicating exactly what's expected of each employee improves motivation.
- **Competence:** Competence is the ability to do something successfully or efficiently. It denotes the sufficiency of knowledge and skills that enable one to act in a variety of situations. To build competence is to create opportunities for others to be challenged, grow, and develop their skills without causing employees to become overly frustrated.
- **Feedback:** Effective feedback shapes performance, builds confidence and encourages sustained effort. Personnel is provided about their performance through feedback.
- **Respect:** Respect is a feeling of deep admiration for someone or something elicited by their abilities, qualities and achievements. Respect affects employee motivation. It makes employees feel a natural part of the team.
- **Rewards:** Rewards are a device to improve motivation among employees. Employee reward systems refer to programs that reward performance and motivate employees on individual or group levels. Rewards may be financial or non-financial. It is essential for the performance and growth of organization.
- **Empowerment:** Empowering is giving authority and liberating the potential of employees. It is a measure designed to increase the degree of autonomy and self-determination in people and communities to enable them to represent their interests in a responsible and self-determined way, acting as their authority.
- **Recognition:** Recognition may refer to something given in recognition of achievement. It is one of the strong motivating factors. Employees feel comfortable when they are praised and recognized. Recognition may be formal or informal. It is a powerful tool to foster a team spirit and encouragement to achieve a high level of productivity.

Factors that motivate the job of employees have to be established after a respectful relationship. Then find out the incentives which are most influential to the staff of the organization concerned. Several theories have been developed to describe the factors that motivate employees to deliver better services. The motivational level of employees varies. But these theories do not point to any specific factors that have a uniform effect. These are the situational realities that make these factors of job motivation relevant.

Factors affecting Job motivation in University Libraries

The composition of job motivation of libraries would vary from university to university and professional to professionals. Six factors of job motivation are identified for the study. The difference in perception of LIS professionals on the factors relating to the motivation in universities in Kerala has collected and analyzed.

Interest in work: Interest in work encourages productive behaviour in an organization. It promotes the performance of employees in the organization. Employees are interested in the work environment to satisfy their needs.

Co-workers: One can motivate the performance of co-workers in a job. It encourages a harmonious atmosphere in the workplace and strengthened professional relationships in the workplace. A creative and friendly working environment acknowledges the employee's achievement.

Promotion and growth: More responsibility and progress in the job motivate employees in work. It usually encourages higher status, authority and responsibility.

Pay: Economic aspects of the employees have an important role in motivating their job in service organisations like libraries. It is related to job satisfaction, output and workability of employees.

Opportunities for higher studies and training: Employees are motivated when they understand their work. Training and higher studies improve the quality of work and outcome in the organization.

Qualification and Advancement: Qualification and advancement significantly influence employees' motivation skills in an organization. It helps to succeed in the job. It is the responsibility of the employer to encourage their employees for improving their qualifications and advancement.

The categorization of factors is fairly exhaustive. In the absence of an agreed list of factors relating to job motivation, the present study has selected the six factors. No motivation theory describes the entire dimensions of job motivation.

Conclusion

Job motivation significantly influences commitment of employees in the organization. It also determines the performance and job satisfaction level of the personnel. Job Motivation is a management tool for manpower planning and the development of human resources in organizations.

References

Arlt, Skemp K.M., & Toupence, R. (2007). The Administrator's Role in Employee Motivation (pp.28-34). Coach & Athletic Director.

Deckers, L. (2010). Motivation; Biological, Psychological and Environmental (3rd ed., pp. 2–3). Boston, MA: Pearson.

Gomes, D. R., Asseiro, V., & Ribeiro, N. (2013). Triggering employee motivation in adverse organizational contexts: "going the extra mile" while holding hands with uncertainty? Business and Management Research. 2 (1),41-54.

Herzberg, F. (1968). One more time: How do you motivate employees? Boston: Harvard Business Review.

Jex, S.M., & Britt, T.W. (2008). Organizational Psychology. Hoboke, New Jersey: John Wiley & Sons, Inc.

Kishore, Jugal. (1981). Personnel Management in Libraries. New Delhi: Ess Ess Publications.

Maslow, A.H. (1943). A Theory of human motivation. Psychological Review,50(4),370–96. doi:10.1037/h0054346

Miao, C. F., & Evans, K. R. (2007). The impact of salesperson motivation on role perceptions and job performance: A cognitive and affective perspective. Journal of Personal Selling and Sales Management, 27 (1),89-101.

Mitchell, T.R., & Daniels, D. (2003). Motivation. Handbook of Psychology. In W.C. Borman, W.C., Ilgen, D.R. & Klimoski, R J(ed.) (Vol.12, pp.225-54) Industrial Organizational Psychology. New York: Wiley.

Pinder, C.C. (2008). Work Motivation in organizational behaviour, (2nd ed.). New york: Psychology press.

Srivastava, S. K., & Barmola, K. C. (2011). Role of motivation in higher productivity. Management. Insight. VII (1),88-99.

5

Data Analysis and Interpretation

Introduction

The most significant part of the research report is analysis and data interpretation. This chapter presents the analysis of data gathered through the questionnaire designed for the study. The study's primary data was collected from 266 (two hundred and sixty-six) Library and Information Science (LIS) professionals working in the eight-state universities in Kerala. It includes a detailed analysis of major results and pertaining information for arriving at a valid conclusion. It helps to interpret the results of the study from the proper perspective. Based on the set forth objectives and hypotheses formulated, the data collected through questionnaires were tabulated, and statistical analyses have been done to formulate the result of the study.

The present chapter is organised around three sections which are according to the nature of the analysis performed. In the following section, data on the various aspects are presented.

I Background factors of LIS professionals.

II Factors of job motivation.

III Total motivation score

In the study, libraries refer to university library/central library, departmental libraries, study centre libraries and constituent college libraries of universities in Kerala. The present study is limited to the libraries of eight major state universities in Kerala. The libraries of the universities are Kerala University Library (KUL), University of Calicut Library, CHMK Library (UoCL), Mahatma Gandhi University, Kerala Library (MGUL), Cochin University of Science and Technology Library (CUSATL), Kannur University Library, Dr. Hermann Gundret Central Library (KAUL), Sree Sankaracharya University of Sanskrit Library (SSUSL), Kerala Agricultural University Library and Information system (KAULIS), Kerala Veterinary and Animal Sciences University Library (KVASUL). The abbreviated form of the name is represented in their respective university libraries.

Background Factors of LIS Professionals

The study investigated the background factors of the LIS professionals working in the universities of Kerala. The background factors may influence the dimension of job motivation. Cadre, age, gender, place of residence, marital status, type of family and salary and dependent variables are various dimensions of job motivation.

Cadre

Human resources have a prominent role in the effective management of any library. There are seven categories of LIS professionals by their cadre in the selected universities of Kerala. The entry cadre of the LIS professionals is Library Assistant or Professional Assistant Grade II. They are subsequently promoted to higher-level posts. In the post of Assistant Librarians and Deputy Librarians, both the UGC/ICAR scale of pay and the state scale of pay. University Librarian heads University Library. In Kerala Agricultural University and Kannur University, there is Central Library which Librarian heads. University Librarian/Librarian is a statutory post.

The details of human resources available in selected university libraries are presented in Table 5.1.

Table 5.1 shows cadre wise distribution of respondents of LIS professionals.

Table 5.1: Cadre-wise Distribution of Respondents

Cadre		Name of Libraries								Total
		KUL	UoCL	MGUL	CUSATL	KAUL	SSUSL	KAULIS	KVASUL	
University Librarian	n	0	0	0	0	0	0	0	0	0
	%	0.00	0.00	0.00	0.00	0.00	0.00	0.00	0.00	0.00
Deputy Librarian	n	3	1	3	3	0	1	1	0	12
	%	4.00	1.67	7.14	8.57	0.00	6.67	10.00	0.00	4.51
Assistant Librarian	n	31	16	7	16	1	0	4	2	77
	%	41.33	26.67	16.67	45.71	4.35	0.00	40.00	33.33	28.95
Reference Assistant (HG)/ Reference officer (HG)	n	0	0	0	0	0	0	4	1	5
	%	0.00	0.00	0.00	0.00	0.00	0.00	40.00	16.67	1.88
Reference Assistant/ Junior Librarian/ Reference officer	n	14	13	10	8	5	2	0	2	54
	%	18.67	21.67	23.81	22.86	21.74	13.33	0.00	33.33	20.30
Technical Assistant/ Professional Assistant Gr. I	n	25	22	15	7	10	5	0	0	84
	%	33.33	36.67	35.71	20.00	43.48	33.33	0.00	0.00	31.58
Professional Assistant Gr-II/ Library Assistant	n	2	8	7	1	7	7	1	1	34
	%	2.67	13.33	16.67	2.86	30.43	46.67	10.00	16.67	12.78
Total	n	75	60	42	35	23	15	10	6	266
	%	100	100	100	100	100	100	100	100	100

Table 5.1 demonstrates that there are seven cadres among the LIS professionals in the universities of Kerala. The data analysis reveals that the entire staff of selected university libraries is 266. University Librarian/Librarian post, which is the university library's head, is lying vacant in all universities. Next to University Librarian is the Deputy Librarian cadre. Respondents working in that cadre are 4.51 per cent, and Assistant Librarians are 28.95 per cent. The total percentage of Reference Assistants (Higher Grade)/Reference officer (Higher Grade) is 1.88 per cent, and Reference Assistant/Junior Librarian/ Reference officer is 20.30 per cent. Respondents in the cadre of Technical Assistant/Professional Assistant Grade. I are 31.58 per cent, and 12.78 per cent of respondents are in the cadre of Professional Assistant Gr-II/Library Assistant. The maximum numbers of respondents are in the cadre of Assistant Librarian followed by Technical Assistant/Professional Assistant Grade. I. From the study, more than 50 per cent of respondents are in the supervisory and above level.

Age

Age was operationally defined as the number of years completed by the respondents at the present investigation. It helps to find out the potential population of the LIS professionals. The respondents were classified into four age groups. They are up to or below 40 years, 41 to 45, age group of 46 to 50 and age group of above 50.

Table 5.2 shows the age-wise distribution of the LIS professionals

Table 5. 2: Age-wise Distribution of LIS professionals

Sl.No.	Age group (In Years)	Frequency	Per cent
1.	Up to or below 40	87	32.7
2.	41 to 45	61	22.9
3.	46 to 50	61	22.9
4.	Above 50	57	21.4
	Total	**266**	**100**

Table 5.2 indicate that 32.7 per cent of respondents are in the age group of up to 40 years. Respondents are in the age group of 41 to 45 years of age, and 46 to 50 years of age group is 22.9 per cent each. Respondents are in the age group above 50 years of age is 21.4 per cent. It may be inferred from the study is that majority of the library staff are below or up to 40 years. It indicates that there is scope for improving their job skills.

Gender

Both male and female LIS professionals were selected as respondents of the study.

Table 5.3 presents the gender-wise distribution of LIS professionals.

Table 5.3: Gender - wise Distribution of LIS professionals

Sl.No.	Gender	Frequency	Per cent
1.	Male	118	44.4
2.	Female	148	55.6
Total		**266**	**100**

Table 5.3 reveals that the number of female respondents is greater than male respondents. Gender is a socio-demographic variable. It is used to find out the status of gender. Gender is an independent variable. It is used to test with other dependent variables selected for the study. The percentage of male respondents is 44.4 per cent, and female respondents are 55.6 per cent. Onuoha., Samuel, and Ojo (2014) opined that library administrators adopt a policy of equal treatment for each gender and review the conditions attached to promotion, among others, as measures for enhancing job satisfaction.

Place of Residence

In India majority of the population is living in rural areas. But the majority of academic institutions in Kerala is located in urban areas. To determine the representation of the LIS professionals from rural, urban and peri-urban backgrounds, place of residence wise distribution of respondents was collected for the study. Respondents were classified into three groups such as urban, rural and peri-urban.

Table 5.4 presents categorization of LIS professionals based on their place of residence.

Table 5.4: Place of Residence-wise Distribution of LIS professionals

Sl.No.	Place of Residence	Frequency	Per cent
1.	Urban	85	31.95
2	Rural	152	57.14
3.	Peri-urban	29	10.9
	Total	**266**	**100**

It is evident from Table 5.4 that number of respondents belonging to the rural background is greater than the number of respondents who belong to the urban background. Respondents with urban backgrounds were 31.95 per cent, while those with rural backgrounds were 57.14 per cent and peri-urban backgrounds were 10.9 per cent. The highest numbers of respondents are from rural backgrounds. Kerala State Library council has a predominant role in rural areas of Kerala with a chain of thousands of libraries. Jharia and Muley (2014) observed that in rural areas, people like farmers, school children, women and youth are taking active part in development through library.

Marital Status

Table 5.5 shows categorization of respondents according to their marital status.

Table 5.5: Marital Status-wise Distribution of Respondents

Sl.No	Marital Status	Frequency	Per cent
1	Married	259	97.4
2.	Unmarried	7	2.6
	Total	**266**	**100**

Respondents of the study are classified as married and unmarried. It is evident from Table 5.5 that the numbers of married LIS professionals are greater than the unmarried LIS professionals. The married LIS professionals are 97.4 per cent, and the unmarried LIS professionals are only 2.6 per cent. Jayaraman and Tamizhchelvan (2015) pointed out that there is no robust correlation between marital status and job satisfaction even it is controlling age and sex.

Type of Family

Table 5.6 presents respondents' categorisation based on the nature of their family, i.e., and they are categorized as the nuclear family and joint family.

Table 5.6: Family-wise Distribution of Respondents

Sl.No.	Nature of Family	Frequency	Per cent
1.	Nuclear	196	73.7
2.	Joint	70	26.3
	Total	**266**	**100**

It is evident from Table 5.6 that number of respondents who belong to a nuclear family is 73.7 per cent, and joint family is only 26.3 per cent.

Salary

Salary is a factor for motivating the work of employees. It is one of the factors for the commitment of the employees. It attracts and retains employees in any organization, including libraries. Respondents are classified as the LIS professionals drawing the UGC/ICAR scale of pay and those LIS professionals who are getting salary in the state scale of pay.

Table 5.7 presents salary wise distribution of respondents.

Table 5.7: Salary-wise Distribution of Respondents

UGC/ICAR Scale of pay (pre-revised) (In Rs.)	Frequency	Per cent
15600—39100	39	84.78
37400-67000	7	15.22
Total	**46**	**100.00**
State Scale of pay (pre-revised) (In Rs.)	**Frequency**	**Per cent**
27800-59400	29	13.18
32300-68700	82	37.27
35700-75600	59	26.82
39500-83000	25	11.36
42500-87000	20	9.09
55350-101400	5	2.27
Total	**220**	**100.00**

It is depicted in Table 5.7 that only 46 respondents were working in the UGC/ICAR scale of pay and the respondents who were working in the state scale of pay is 220. The majority of the LIS professionals were working in the state scale of pay. There is no recruitment in the post of UGC/ICAR scale of pay in the cadre of LIS professionals. Library and information professionals are appointed only through state scale of pay.

Membership in Library Associations

Library associations serve as the voice of the profession and professionals. Most of the respondents selected for the study have membership in professional library associations to promote librarianship as a profession. The LIS professionals in Kerala have membership in Kerala Library Association, Indian Library Association (ILA), KELPRO, Indian Association of Special Libraries and Information Centres (IASLIC), etc.

Table 5.8 presents details of Membership of library professionals in professional library associations.

Table 5.8: Membership in Professional Bodies–wise Distribution of Respondents

Sl.No.	Name of Professional Associations*	Frequency	Per cent
1.	Kerala Library Association (KLA)	162	88.04
2.	KELPRO	47	25.54
3.	Indian Library Association (ILA)	20	10.87
4.	Indian Association of Special Libraries and Information Centres (IASLIC)	11	5.98
5.	Others	21	11.41
6.	No Membership	77	28.90
*Multiple response			

Table 5.8 reveals that the highest number of respondents (88.04%) who have

membership is Kerala Library Association (KLA), followed by KELPRO (25.54%). It is also indicating that 28.90% of respondents have no membership in any professional library association.

Factors of Job Motivation

The concept of job motivation is applicable to non-profit organization like university libraries, where groups of people gather for a common purpose of their academic interest. These attitudes can affect the satisfaction of personnel with organization. It has an impact on the motivation of employees. Motivation in an organization has to play a crucial role in enhancing the performance of personnel. Motivation includes both intrinsic and extrinsic motivation. The following are factors studied to assess the job motivation of LIS professionals working in libraries.

The following six factors are studied to assess the job motivation of the LIS professionals working in the libraries of universities. The factors relating to the job motivation listed below have statements. These factors regarding job motivation have influence and motivate employees. The relevant factors are: --

I Interest in work

II Co-workers

III Promotion and growth

IV Good pay

V Opportunities for higher studies and training

VI Qualification and advancement

Likert's five-point scale was used for measuring the perception of job motivation of the LIS professionals. The six factors of job motivation were identified. Several positive and negative statements relating to each factor are used to elicit the opinion of the LIS professionals relating to their job motivation in the university libraries. The respondents of the study were asked to mark their opinion on a five-point scale consisting of each statement. Strongly Agree (SA), Agree(A), Neither Agree nor Disagree (NAD), Disagree(D), Strongly Disagree (SD) are the option given to the respondents. The scale item carries values 5,4,3,2, and1, respectively. The total score of each statement under each factor was calculated with its respective values. The total score secured by each statement shows its significance in relation to the other statements included under each factor.

Interest in Work

The perceived value of work, work climate, work-life balance and satisfaction with relationships in life are major factors that shape job and employee perceptions of overall quality of work. Adequate job security and pride in doing the job will motivate LIS professionals. It will encourage and executing work effectively.

Table 5.9 depicts interest in work among the LIS professionals.

Table 5. 9: Interest in Work among LIS Professionals

Statements	Mean	SD	t	Sig.
I take pride in doing my job as well as I can.	4.19	0.64	-30.266	0.000
I try to think of ways of doing my job not effectively.	2.55	0.89	8.375	0.000
My opinion of myself does not go down when I do my job badly.	2.78	0.87	4.165	0.000
I feel unhappy when my work is not up to my usual standard.	3.73	0.80	-14.917	0.000
I like to look back on the day's work with a sense of satisfaction.	3.80	0.66	-19.873	0.000
There is no adequate job security in my job.	2.34	0.87	12.517	0.000

Table 5.9 indicates that in the opinion score of the statement that *I take pride in doing my job as well as I can* is found to be the highest score of 4.19. It is significantly higher than the mean of the response scale as the significance level of the t-test is less than 0.05. The LIS professionals can take pride in doing their job as well as they can. The second important aspect of interest in work is the statement *that I like to look back on the day's work with a sense of satisfaction.* The mean score of that aspect is 3.80. The other important aspect of interest in work is the statement *I feel unhappy when my work is not up to my usual standard.* The mean score of that aspect is 3.73. The other aspects of interest in work in descending level are *My opinion of myself does not go down when I do my job badly* (mean score 2.78), and *I try to think of ways of doing my job not effectively* (mean score 2.55). The lowest mean score is found for *There is no adequate job security in my job* (mean score 2.34). It is significantly lower than the mean of the response scale. It means there is an insignificant opinion about adequate security about the job.

Co-Workers

Work in a library and information centre is teamwork. Technically proficient and socially supportive co-workers are assets to the fellow workers of the library. A caring and cooperative environment in organization encourage social harmony and support in the workplace. It will reduce boredom and undue work pressure and encourages better understanding among the professionals.

Table 5.10 depicts the opinion of library professionals about their co-workers.

Table 5. 10: Opinion of LIS Professionals about co-workers

Statements	Mean	SD	t	Sig.
My fellow workers are technically proficient and socially supportive.	3.89	0.70	-20.611	0.000
Conflicts are not constructively and positively resolved among the staff.	3.03	0.94	-0.586	0.558
There is caring and cooperative environment in our organisation.	3.62	0.81	-12.619	0.000
I am frequently put under undue work pressure by my colleagues.	2.50	0.78	10.557	0.000
There is lack of understanding among employees of our organisation.	2.78	0.95	3.742	0.000
There is social harmony and respect in our organi-sation.	3.70	0.84	-13.615	0.000

It is evident from Table 5.10 that in the component of a co-worker in the opinion score of the statement *"my fellow workers are technically proficient and socially supportive"* is found to be the highest with a mean score of 3.89. It is significantly higher than the mean of the response scale as the significance level of the t-test is less than 0.05. It reveals that the co-workers are technically proficient and socially supportive. The second important aspect of co-workers is the statement that *There is social harmony and respect in our organisation.* The mean score of that aspect is 3.70. The other important aspects of co-workers in descending level are in the *caring and cooperative environment in our organisation* (mean score 3.62), and *Conflicts are not constructively and positively resolved among the staff* (mean score 3.03). The other important aspects of co-workers are on the statement of *lack of understanding among employees of our organisation* (mean score 2.78). It is significantly lower than the mean of the response scale. The lowest mean score is found for *I am frequently put under undue work pressure by my colleagues* (mean score 2.50). It indicates that there is no excessive work pressure from colleagues. Ahmad et al. (2010) reported that satisfaction of employees with organizational climate enhances positive organizational outcomes such as efficiency, productivity, organizational commitment and cohesiveness of co-workers.

Promotion and Growth

Status, salary, service conditions, recruitment and promotional policy, and managerial relations are the promotional factors encouraging job motivation. Promotional factors of job lead to increased productivity. In a library, professionals can develop their career to attain a job career and helps them to reach their personal and professional goals.

Table 5.11 presents the promotion and growth among library and information science professionals.

Table 5.11: Promotion and Growth among LIS Professionals

Statements	Mean	SD	t	Sig.
I can develop my career in my organisation.	3.64	0.78	-13.268	0.000
I am not satisfied with the criteria for promotion in my organisation.	3.05	1.01	-0.792	0.429
In our organisation there is opportunity for ultimate achievement in my job career.	3.11	0.95	-1.815	0.071
I have opportunity to achieve personal and professional goal.	3.34	0.90	-6.211	0.000
There is no scope of new ideas in my job encouraged by our organisation.	2.71	0.90	5.206	0.000
I would have got more promotions if I were in the ministerial cadre of this organisation.	3.17	1.03	-2.631	0.009

Table 5.11 indicates that in the opinion score of the statement, *I can develop my career in my organisation*, which is the highest with a mean score of 3.64. It is significantly higher than the mean of the response scale as the significance level of the t-test is less than 0.05. It means that LIS professionals generally agree that they can develop their careers in their organisation. From the result, it can be inferred that LIS professionals can build their careers in their organisation in promotion and growth. The second important aspect of promotion and growth is on the opinion score statement *I have opportunity to achieve a personal and professional goal.* The mean score of that aspect is 3.34. The other important aspects of promotion and growth in the order of descending level are *I would have got more promotions if I were in the ministerial cadre of this organisation* (mean score 3.17*), In our organisation, there is opportunity for ultimate achievement in my job career* (mean score 3.11), *and I am not satisfied with the criteria for promotion in my organisation* (mean score 3.05). The mean scores of these aspects are significantly higher than the mean of the response scale. The lowest score is found for *There is no scope of new ideas in my job encouraged by our organisation* (mean score 2.71). The result indicates that there is no scope for new ideas in the job of the LIS professionals encouraged by the organisation of library professionals in the aspects of promotion and growth.

Good Pay

Library staff derives satisfaction from their salary. So, salary should be commensurate with the duties and responsibilities of employees. It should be competitive to similar organization and similar professions. Good pay should motivate them to continue in the library profession.

Table 5.12 depicts opinions about good pay among the LIS professionals.

Table 5.12: Opinion about Good Pay among LIS Professionals

Statements	Mean	SD	t	Sig.
Salaries are not commensurate with responsibility of the employees.	2.92	0.92	1.460	0.145
Pay scale of universities in library staff is not competitive with similar organisations.	3.43	0.99	-7.101	0.000
I prefer my job with a different designation although pay is same.	3.06	0.91	-1.079	0.282
Salary is not an important factor motivating me at work in this profession.	3.18	0.92	-3.253	0.001
I will continue in the library profession regardless of the pay received.	3.49	0.83	-9.627	0.000
Salary of library professionals is as fair as that of other categories of staff.	2.91	1.06	1.393	0.165

Table 5.12 indicate that in the element of good pay, the mean score of the statement *Salaries is not commensurate with the responsibility of the employees is* 2.92 as the significance level of the t-test is greater than 0.05. It is lower than the mean of the response scale. It means that LIS professionals agree that salaries are not commensurate with responsibility. From the result, it can be seen that salaries not commensurate with responsibility are below average. The opinion score of the statement that *I will continue in the library profession regardless of the pay received* found to be the highest score of 3.49. It indicates that the LIS professionals will continue to the profession irrespective of the payment received. The second aspect of the good pay is on the statement of *Payscale of universities in library staff is not competitive with similar organisations*. The mean score of that aspect is 3.43. The other important aspect of good pay in order of descending level are *Salary is not an important factor motivating me at work in this profession* (mean score 3.18), and *I prefer my job with a different designation although the pay is the same* (mean score 3.06). The lowest score is found for *Salary of library professionals is as fair as that of other categories of staff* (mean score 2.91). It is lower than the mean of the response scale. The study indicates that LIS professionals disagree that salary of LIS professionals is as fair as that of other categories of staff.

Opportunities for Higher Studies and Training

Opportunities for higher studies and training affect the prospect of professional development of library staff positively. As qualified staff decrease, training becomes more important. Training improves efficiency and familiarity with the latest development with modern technology. So, organizations should encourage librarians to professional training, seminars, workshops and conferences. The study states, along with salary for attaining higher qualification will improve the efficiency of LIS professionals.

Table 5.13 depicts opportunities for higher studies and training among LIS professionals

Table 5. 13: Opportunities for Higher Studies and Training among LIS Professionals

Statements	Mean	SD	t	Sig.
My organisation provides study leave along with salary to attaining higher qualifications.	2.35	0.96	11.114	0.000
I am not satisfied with duty leave and incentives to attend conferences seminar etc.	3.14	1.10	-2.011	0.045
My organisation does not provide regular training in latest developments.	3.26	1.03	-4.179	0.000
The training I received from my organisation is not of high quality.	3.09	0.91	-1.680	0.094
I have given opportunity to implement what I have learnt from the training programmes.	3.17	0.90	-3.011	0.003
Organisation encourages librarians for attending professional training, seminars etc.	2.97	1.04	0.532	0.595

Table 5.13 reveals that in the element of "opportunities for higher studies and training", the lowest mean score of the statement "my organization *provides study leave along with salary for attaining higher qualifications"*. The average mean score of the statement is 2.35. It is significantly lower than the mean of the response scale as the significance level of the t-test is greater than 0.05. It indicates that in the matter of opportunities for higher studies and training, LIS professionals agree that no study leave will be provided along with the salary for attaining higher qualification. The opinion score of the statement " *my organisation does not provide regular training in latest developments*" found to be the highest mean score of 3.26. It is greater than the mean of the response scale. It indicates that regular training in the latest developments does not provide to LIS professionals. The other important aspects of opportunities for higher studies and training in order of the descending level are *I have given opportunity to implement what I have learnt from the training programmes* (mean score 3.17), *I Not satisfied with duty leave and incentives to attend conferences seminar etc.* (mean score 3.14) and *The training I received from my organisation is not of high quality (mean score 3.09).* The other important

aspect is on the statement of *Organisation encourages librarians for attending professional training, seminars etc.*(mean score 2.97) which is lower than the mean of the response scale. Akintunde (2006) reported the need for librarians to undergo compulsory routine retraining to pre-empt technological obsolescence and to cope with the unfolding exigencies of the technological age emphasized.

Qualification and Advancement

Qualification is coupled with both personal and professional aspirations. Library professionals are fully qualified to handle their present job because of their qualification. They have to possess the professional qualification to enter into the field of librarianship. There should be opportunity for future career advancement for library staff. Qualifications should motivate to get adequate promotion and higher pay scales.

Table 5.14 depicts opinions about qualification and advancement among LIS professionals.

Table 5.14: Qualification and Advancement among LIS Professionals

Statements	Mean	SD	t	Sig.
I am fully qualified to handle my present job.	3.96	0.76	-20.615	0.000
Career advancement scheme is taken seriously in our Organisation.	2.98	0.96	0.385	0.700
I have never got opportunity to see my appraisal report and discuss it with my superior.	3.15	0.78	-3.065	0.002
Qualification of staff in library does not reflect merit properly.	3.12	0.85	-2.304	0.022
My promotional opportunities are not linked with qualification.	3.21	0.98	-3.434	0.001
Qualifications motivate to get sufficient promotion and higher pay scales.	3.40	0.94	-6.953	0.000

Table 5.14 depict that, in the element of qualification and advancement, the mean score of the statement *I am fully qualified to handle my present job* found to be the highest score of 3.96. It is significantly higher than the mean of the response scale as the significance level of the t-test is lower than 0.05. It indicates that LIS professionals agree that they are qualified enough to handle their present job. From the result, it can be inferred that LIS professionals are fully qualified. The second important aspect of qualification and advancement is on the statement *Qualifications motivate to get sufficient promotion and higher pay scales.* The mean score of that aspect is 3.40.

The other important aspects of qualification and advancement in order of descending level are *My promotional opportunities are not linked with qualification* (mean score 3.21), *I have never got opportunity to see my appraisal report and discuss it with my superior* (mean score 3.15) and *Qualification of staff in library does not reflect merit properly* (mean score 3.12). The lowest score is found for *the Career advancement scheme is taken seriously in our Organisation* (mean score2.98). It is lower than the mean of the response scale. The study indicates that the LIS professionals disagree that the career advancement scheme is taken seriously in their organisation. Nguyen (2018) emphasized that promotion focus significantly moderated the over qualification intrinsic motivation relationship.

Total Motivation Score

Table 5.15 depicts the total motivation score of the components of job motivation.

Table 5.15: Total Motivation Score

Statistics	Interest in work	Co-workers	Promotion and growth	Good pay	Opportunities for higher studies and training	Qualification and advancement	Motivation Total
Mean	3.23	3.25	3.17	3.17	3.00	3.30	3.19
Median	3.17	3.17	3.17	3.17	3.00	3.33	3.14
Standard Deviation	0.34	0.38	0.37	0.46	0.36	0.39	0.24
Minimum	2.33	2.33	2.33	2.00	1.83	1.00	2.64
Maximum	4.17	5.00	4.33	5.00	4.17	5.00	4.36

Figure 5.15 indicates the motivation score of each factor selected for the study.

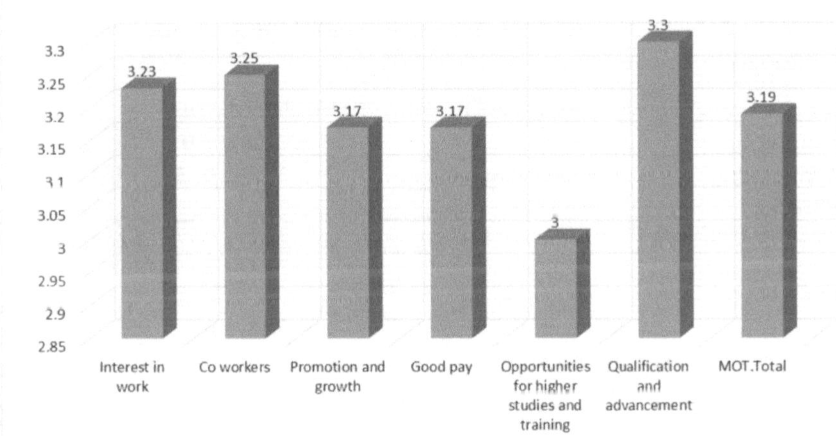

Fig. 5.1: Total Motivation Score

To find out the composite variable representing the job motivation, the average score of its dimensions was computed, and the descriptive statistics of this variable are given below. The total motivation of all the mean variables is 3.19, with a standard deviation of 0.24. The median is found to be 3.14, which is a little smaller than that of mean. The total motivation score of the variable ranges from a minimum of 2.64 to a maximum of 4.36. The total motivation score is presented in figure 5.1. Liunome (2018) emphasized that organizational climate and organizational culture have influences on work motivation of employees and support the goal of institution.

Role of Authorities in Motivating LIS Professionals

The motivation of authority energizes professionals to overcome setbacks in the course of their employment. Organizational structures, resources, processes, climate and culture, and organizational feedback about performance contribute to the motivational processes of library professionals. Both intrinsic and extrinsic factors are the target for effective employee motivation. Motivation is an individual's degree of willingness to exert and maintain an effort towards organizational goals. The study assesses motivational status and factors affecting library professionals of eight university libraries of Kerala. Authorities of universities have to play a significant role in motivating Library professionals for their professional development and carry out their duties and responsibilities. The motivation of university authorities will bring a favourable working environment to the library professionals. Effective cooperation between the university administration and library professionals is a key factor in providing better service to library clientele with their teaching, research and learning conditions.

Table 5.16 depicts the role of university authorities in motivating LIS professionals.

Table 5. 16: Role of University Authorities in Motivating LIS Professionals

Statements	Mean	SD	t	Sig.
Authorities are too tolerant of those who perform poorly in our library.	3.18	0.85	3.464	0.001
Rewards for good work of staff.	3.27	0.77	5.816	0.000
I am free to express my views and suggestion in decision making process.	3.54	0.83	10.576	0.000
Opportunity to learn new skills and technology,	3.65	0.69	15.461	0.000
Opportunity for career development.	3.26	0.71	5.925	0.000

Table 5.16 revealed that in the matter of the role of authorities in motivating LIS professionals, the mean score of the statement *Authorities is too tolerant of those who perform poorly in our library* found to the higher score of 3.18.

It is significantly higher than the mean of the response scale as the significance level of the t-test is greater than 0.05. It indicates that LIS professionals agree that authorities are too tolerant of those who perform poorly in our library. The opinion score of the statement that *Rewards for good work of staff is* found to be the highest mean score of 3.27. It indicates that LIS professionals are rewarded for good work. It is significantly higher than the mean of the response scale as the significance level of the t-test is greater than 0.05. The next important aspect is the role of authorities in motivating LIS professionals in the university libraries in Kerala is on the statement of *"I am free to express my views and suggestion in the decision-making process"*. The mean score of that statement is 3.54. It is significantly higher than the mean of the response scale as the significance level of the t-test is greater than 0.05. The role of authorities in motivating LIS professionals for the statement, *Opportunity to learn new skills and technology*, with a highest mean score of 3.65. It is significantly higher than the mean of the response scale as the significance level of the t-test is greater than 0.05.

The other aspect of role of authorities in motivating LIS professionals in the university libraries in Kerala for the statement of *Opportunity for career development*, with a mean score of 3.26. It is significantly higher than the mean of the response scale as the significance level of the t-test is greater than 0.05. It indicates that library authorities are playing a greater role in motivating the LIS professionals. Yaminfirooz et al. (2013) declared that library management and authorities have an important role in improving organizational climate of libraries and they could provide management strategies to raise interest, creativity and motivation among employees.

Role of Professional Library Associations in Motivating LIS Professionals

Joining a professional organization can open several doors and boost one's career. Professional library associations influence librarian's jobs and their professional career. It helps to develop, maintaining and sustaining professionalism that underpins the profession of librarianship. Professional Associations helps to improve user satisfaction and service outcomes through publications, seminar, conferences, and workshop. It is also motivating library professionals.

Table 5.17 presents an opinion about the role of professional library associations in motivating library and information science professionals.

Table 5.17: Opinion of LIS Professionals about Role of Professional Library Associations

Statements	Mean	SD	t	Sig.
Library Associations do not organise programmes for motivating professionals.	2.93	1.08	1.140	0.255
Programmes of library associations did not motivate my job.	2.94	0.97	1.012	0.312
Library associations provide training in only latest developments like Information Technology etc.	3.05	0.76	-1.051	0.294
We do not get opportunities to attend activities sponsored by library associations.	3.06	0.91	-1.144	0.254
Professional Associations helps to improve user satisfaction and service outcomes through publications.	3.28	0.86	-5.377	0.000
I have attended the programmes of library professional associations in motivating my profession.	3.38	0.86	-7.159	0.000

From Table 5.17, it is clear that in the opinion score of the statement that *Library Associations do not organise programmes for motivating professionals* is 2.93. It is significantly lower than the mean of the response scale as the significance level of the t-test is lower than 0.05. It indicates that LIS professionals disagree that Library associations do not organise programmes for motivating professionals. The opinion score for the statement that *I have attended the programmes of library professional associations in motivating my profession* found to be the highest score of 3.28. It indicates that library professionals are attended the programmes of library professional associations in motivating their profession.

The other important aspect of the role of professional library association in motivating LIS professionals in order of descending level are *Professional Associations helps to improve user satisfaction and service outcomes through publications* (mean score 3.28), *We do not get opportunities to attend activities sponsored by the library associations*, (mean score 3.06). *Library associations provide training in only the latest developments like Information Technology etc.* (mean score3.05). It is significantly higher than the mean of the response scale. The mean score found for *Programmes of library associations did not motivate my job* is 2.94 and is less than the mean of the response scale. It indicates that library professionals disagree that programmes of library associations did not motivate their job. Henczel and Macauley (2013) revealed that library associations have roles and responsibilities in developing, maintaining and sustaining professionalism that underpins the profession of librarianship within its changing environment.

Conclusion

Job motivation related to all factors is examined in this study. Job motivation related to all factors is different in the university libraries of Kerala. The overall analysis of the data in the study revealed that there is the motivation among library and information science professionals in the university libraries to carry out duties and responsibilities. The study shows that the total motivation score of all the factors is above the mean score. The study statistically established that qualification and advancement are ranked first, followed by co-workers among the major factors of job motivation of LIS professionals in the selected universities of Kerala. The study also indicates that library authorities are playing a greater role in motivating the LIS professionals. Library Associations have a significant role in motivating the LIS professionals in their job. To improve the functioning of the university libraries, policy should be formulated by giving emphasise to the factors as mentioned above relating to job motivation.

References

Ahmad, Zulfqar., Ahmad, Zafar., Ahmad, Ashfaq, Nawaz, & Musarrat, Muhammad (2010). Organizational Climate (OC) as Employees' Satisfier: Empirical Evidence from Pharmaceutical Sector. International Journal of Business and Management, 5(10), 214-222.

Akintunde, S.A. (2006). State of ICTs in tertiary institutions in Nigeria: Window on the universities. A paper presented at the NLA 44th Annual National Conference/AGM held at National Centre for Women Development, Abuja, June 18-23, 123-137.

Henczel, S. , & Macauley, P. (2013, 20-23 October,).Professionalism:Exploring the role and responsibility of our professional associations. LIANZA Conference. Hamilton, New Zealand: Wai-Ora, Wai-Maori, Waikato.http://dx.doi.org/10.4236/oalib.1100804

Jayaraman, Irudayam & Tamizhchelvan, Mottaiachettiar. (2015). An empirical Analysis of the relationship between marital status and job satisfaction of Library professionals. International Journal of Applied Engineering Research10(8),21383-21391.

Jharia, Babita., & Muley, Anjali. (2014). Role of library in rural development. International Journal of Computer Science and Network (online),152-157.Retrieved from http://ijcsn.org/RDFCORDGE-2014/C-044.pdf

Nguyen,A.D.(2018).Perceived overqualification and withdrawal among seasonal workers: Would work motivation make a difference? (10746206). Available from ProQuest Dissertations & Theses Global A&I: The Sciences and Engineering Collection. (2042942164). Retrieved from https://search.proquest.com/docview/2042942164?accountid=62114

Onuoha, Uloma Doris., Samuel, Toyosi Daniel., & Ojo, Adebowale Ifeoluwa. (2014). Gender and job satisfaction: A study of library staff in Private University Libraries in Ogun State, Nigeria. Journal of New Media and Mass Communication 1(1), 12-19.

Yaminfirooz, M., Yaminifirouz., & Mousavi, A. (2013). Assessment of organizational climate of libraries at Babol University of Medical Sciences in 2012. Journal of Health Administration (JHA),16 (51),60-69

6

Findings and Conclusion

Any organization which is in the process of scaling heights should find job motivation for its personnel. The job motivation of personnel in any organization should support the personnel and improve productivity and services. The chapter gives accounts of the significant findings of the study, followed by tenability of hypotheses and further areas of research in the related areas, suggestions and conclusion. The present study has been undertaken to find out the job motivation of LIS professionals working in Kerala's major eight state universities. The data of the study were collected from 266 (two hundred and sixty-six) Library and Information Science professionals. The study's objectives are examined with reference to the analysis of data gathered through questionnaires and realized. The result of the study was analysed using MS Excel and SPSS and formulated findings and concluded.

Findings of the Study

The major findings evolved out of the study are summarised and particularised in the following headings.

- To identify the factors that influence the LIS professional's job motivation working the university libraries of Kerala.

  Six factors are considered for the job motivation of LIS professionals working in the universities of Kerala. The factors that influence the job motivation of LIS professionals in the university libraries are interest in work, co-workers, promotion and growth, good pay, opportunities for higher studies and training, qualification and advancement.

- To study the job motivation of LIS professionals in the university libraries of Kerala.

It is interesting to note from the study that most respondents are proud of their work as per their response in relation to their level of agreement to the various aspects of employees' motivation.

Analysis of the elements co-workers shows that most LIS professionals agree that their fellow workers are technically proficient and socially supportive.

In the element of "promotion and growth", the majority of the LIS professionals agree that they can develop their careers in their university.

With regard to the pay of the LIS professionals, it is noticed that LIS professionals will continue the profession regardless of the pay they received.

In "opportunities for higher studies and training", most respondents agree that their organization does not provide regular training in the latest developments. There is little possibility for providing study leave and salary to attaining higher qualifications.

In qualification and advancement, LIS professionals agree that they are fully qualified to handle their present job. The career advancement of LIS professionals is not taken seriously in their universities.

- To find out the total motivation score of all the factors selected for job motivation in the study.

  All the six factors selected for the job motivation of LIS professionals working in the selected libraries of the university are appropriately analyzed. An analysis of all the mean variables' total motivation in the factors of job motivation mentioned in the study is 3.19 with a standard deviation of 0.24.

- To rank the factors which affect the job motivation of LIS professionals in the universities of Kerala.

  It is found in the study is that qualification and advancement followed by coworkers are the most important factor affecting the job motivation of LIS professionals in the selected universities of Kerala.

- To examine the role of authorities in motivating the job of LIS professionals of university libraries in Kerala.

  From the study, there is no significant variation in the opinion of library and information science professionals about the role of university authorities in motivating their job. It means that all the LIS professionals believe that university authorities have a significant role in motivating the job of library and information science professionals.

- To study the role of Library Associations in motivating the job of qualified LIS professionals.

  It has been observed from the study is that there is a significant agreement, amongst the qualified LIS professionals, in the level of agreement in the important role of library associations in motivating their job. Library associations are significantly higher to put substantial emphasis on professional development.

Testing of Hypotheses

The hypotheses in the study are examined with reference to the analysis of data gathered through a questionnaire and tested. The results of the testing of hypotheses are presented below.

H1: There is a lack of motivation among LIS professionals to carry out their duties and responsibilities.

To test this hypothesis, questions were asked in the questionnaire to obtain the data regarding the motivation of LIS professionals in the university libraries of Kerala. The following factors are assessed to study motivation among the LIS professionals in the university libraries of Kerala to carry out their duties and responsibilities. The related factors are interest in work, co-workers, promotion and growth, good pay, opportunities for higher studies and training, qualification and advancement. The composite variable representing job motivation was found out. The total motivation of the entire mean variable is analysed. It reveals that there is above an average level of motivation among the LIS professionals in the university libraries of Kerala to carry out their duties and responsibilities.

Thus, the result fully rejected the null hypothesis H1 that there is a lack of motivation among library and information science professionals in the university libraries in Kerala to carry out duties and responsibilities and accept the alternative hypothesis that there is the adequate motivation among library professionals in the university libraries of Kerala to carry out duties and responsibilities.

H2: Authorities are adequately motivating the job of LIS professionals in the universities of Kerala.

As can be seen from the analysis of data presented, it can be inferred from the study is that library authorities are playing a greater role in motivating the library and information science professionals. From the statements are selected to study, it is indicated that there is significant agreement among the library professionals that there is a greater role from the authorities of the universities in motivating library professionals. All the library professionals have the opinion that university authorities have a significant role in motivating library professionals. Findings prove this.

Hence the result of the study accepts the hypothesis that the authorities of the universities play a significant role in motivating library professionals in the university libraries in Kerala.

H3: Role of professional Library Association is minimum in motivating the job of LIS Professionals.

Regarding the role of professional library associations in motivating library professionals, LIS professionals have membership in professional Library Associations in all the selected universities. Professional library association in motivating library professionals is found. The mean score is found to be average of the response scale or above average. The result indicates that library associations have a moderate or above moderate role in motivating library professionals of different characteristics. Findings prove this.

Hence the result rejects the null hypothesis that professional library associations have a minimal role in motivating the library professionals of university libraries of Kerala and accepts the alternative hypothesis that professional library associations have a significant role in motivating the library professionals of university libraries of Kerala.

Suggestions

- To improve perception of employees of the libraries towards the factors relating to the job motivation
- Motivate the library professionals with more incentives and good salary packages in academic libraries.
- LIS professionals should be duly promoted and reasonably recognized in a good working environment. It creates a feeling of motivation in the job performed.
- Library Associations must be a platform for motivating the LIS professionals for the improvement of library services.
- The Universities should also introduce Career Advancement Programmes to do Ph.D., and Post-doctoral programmes to Library and Information professionals.
- Higher authorities should regularly touch with all the LIS professionals and give continuous training programmes on the latest development in the field.
- The attitude from the authorities should rise to the occasion towards the essentials of the library for the improved for the job motivation.

- There is a need for delegation of powers to the supervisory level to make decisions and solve problems to encourage decentralized decision-making.
- The introduction of training and social outreach programmes in the library and information centre will introduce new vigour to the qualified LIS professionals.
- A concerted effort from all the stakeholders, to improve job motivation of LIS professionals working in the university libraries.

Areas of Further Research

The study conducted in one area throws light on other issues. In the process of conducting this study, some connected issues deserving of depth study has been identified. The areas suggested for undertaking similar studies are suggested. Based on the present study, the following research problems are suggested for further investigation.

- The present study was undertaken in the eight selected state universities of Kerala. Replicated investigations in the same universities and other libraries, such as academic libraries, special libraries, and public libraries, help better understand job motivation LIS professionals of other libraries and information centres.
- In the wake of technological innovation and advancements, the impact on the job motivation of library and information professionals in the library and information centres has to be documented.
- Larger samples and the nature of library and information centres and new variables should be undertaken to study job motivation.
- There are several interdisciplinary points for conceptualizing, constructing and interpreting the present study for better employee motivation in library and information centres.

Conclusion

Job motivation has an impact on libraries and information centres. The study was conducted mainly to study job motivation in the university libraries of Kerala. The objectives and hypotheses of the study have been fulfilled, and observations are presented. Factors of job motivation, namely interest in work, co-workers, promotion and growth, good pay, opportunities for higher studies and training, qualification and training along with the role of authorities in motivating LIS professionals and role of library associations in motivating

LIS professionals prevailing in these university libraries are also found out. The study will set a new benchmark and help scholars, academicians, library and information science professionals to enhance their performance. It will open a window on an emerging trend in the importance of university libraries in the new learning environment. Thus, it is concluded from the study is that improvement in all the aspects would lead to job motivation of LIS professionals in the university libraries of Kerala. Finally, there is a need for cooperation among all the university stakeholders for the success of job motivation among the LIS professionals of university libraries.

References

Abifarin, Abimbola. (1997). Motivating staff in Nigerian university libraries. Library Management, 18(3), 124 –128.

Agarwal, U. N. (1987). Measuring job-involvement in India. Indian Journal of Industrial Relations, 14, 219-31.

Ahmad, Zulfqar., Ahmad, Zafar., Ahmad, Ashfaq, Nawaz, & Musarrat, Muhammad. (2010). Organizational Climate (OC) as Employees' Satisfier: Empirical Evidence from Pharmaceutical Sector. International Journal of Business and Management, 5(10), 214-222.

Ahmed, Ishfaq., Muhammad Musarrat Nawaz., Naveed Iqbal., Imran Ali., Zeeshan Shaukat., & Ahmad Usman. (2010). Effects of motivational factors on employees job satisfaction: A case study of University of the Punjab, Pakistan. International Journal of Business and Management, 5(3), 70-80. doi: http://dx.doi.org/10.5539/ijbm.v5n3p70

Aina, L. O. (2004). Library and Information Science Text for Africa (pp.281-282). Ibadan: Third World Information Service Ltd.

Akintunde, S.A. (2006). State of ICTs in tertiary institutions in Nigeria: Window on the universities. A paper presented at the NLA 44th Annual National Conference/AGM held at National Centre for Women Development, Abuja, June 18-23, 123-137.

Alajmi,Bibi., & Alasousi, Hessah. (2018). Understanding and motivating academic library employees: Theoretical implications. Library Management. https://doi.org/10.1108/LM-10-2017-0111

Ali Al-Aufi, Khulood Ahmed Al-Kalbani. (2014). Assessing work motivation for academic librarians in Oman. Library Management, 35(3), 199 – 212.https://doi.org/10.1108/LM-03-2013-0020

Alqudsi-ghabra., Taghreed, Mansouri., & Huda H. (2010). Staff motivation at Kuwait University Libraries. Journal of Information & Knowledge Management 9(4), 355-363.

American Psychological Association. (2010). Publication manual of the American Psychological Association (6th ed.). Washington, DC: Author.

Amune, Juliana Bosede. (2013). Job Motivation as a predictor of job satisfaction among professional and non-professional library staff in Ambrose Alli University, Ekpoma. International Journal of Innovative Research & Development, 2(5), 1477-1497.

Antwi, I.K., & Bello, M.A. (1993). Motivation and productivity in academic libraries: A case study of the Abubakar Tafawa Balewa University Library. Bauchi Nigeria, Library Management, 14(5), 27-34.

Arlt, Skemp K.M., & Toupence, R. (2007). The Administrator's Role in Employee Motivation

(pp.28-34). Coach & Athletic Director.
Asante, Job; Yahaya, Wahabu & Yaahowaah, F. (2020). Motivating Employees using Equitable Pay Systems in Ghana: The Single Spine Salary Structure in Perspective. Asian Journal of Management.11(3),321-328.
Aziagba, Philip Chime. (2009). Staff development as motivation for paraprofessionals: A case study in southern Nigeria. Chinese Librarianship: An International Electronic Journal, (29), 1-8. Retrieved from http://www.iclc.us/cliej/cl27aziagba.pdf
Babalola, G. A., & Nwalo, Kin. (2013). Influence of job motivation on the productivity of librarians in Colleges of Education in Nigeria. Information and Knowledge Management, 3(5), 70-75.
Badu, Edwin Ellis. (2005). Employee motivation in University Libraries in Ghana: A comparative analysis. Library and Information Sciences, 21(1), 38-46.
Bamgbose, Adeoye Augustine., & Ladipo, Sunday Olusola. (2017). Influence of motivation on academic library employees' performance and productivity in Lagos, Nigeria. Journal of Information and Knowledge Management., 8 (2), 33 – 47.
Baro, Emmanuel E. (2012). Job rotation program evaluation: The Niger Delta University Library. Aslib Proceedings, 64(4), 388 - 404. doi: http://dx.doi.org/10.1108/00012531211244743
Bernstein, A. M. (2011). A Herzbergian look at academic librarians and job satisfaction. Georgia Library Quarterly, 48(3), 6-21.
Boers, G. G. A. (1997). Mentorship and motivation: Librarians and general staff working together. Georgia Librarian, 34(4), 9-1.
Crawford, A. (2017). The Meaning of the Library: A Cultural History. Princeton: Princeton University Press.
Deckers, L. (2010). Motivation; Biological, Psychological and Environmental (3rd ed., pp. 2–3). Boston, MA: Pearson.
Edem, Selong U. (1994). Quantitative measurement of publication output among librarians in Nigerian Universities. Annals of Library Science and Documentation, 41(2), 55-62.
Gairola, S. C. (2015). Assessment of Job Performance of Forest Guards in India: Relationship with Job Motivation, Satisfaction, Stress and other Critical Factors. Indian Forester, 141(3), 235-246.
Gomes, D. R., Asseiro, V., & Ribeiro, N. (2013). Triggering employee motivation in adverse organizational contexts: "going the extra mile" while holding hands with uncertainty? Business and Management Research. 2 (1),41-54.
Hackman, J.R., & Lawler, E.E. III. (1971). Employee reactions to job characteristics, Journal of Applied Psychology, 55 (3), 259-286.
Henczel, S., & Macauley, P. (2013, 20-23 October,). Professionalism: Exploring the role and responsibility of our professional associations. LIANZA Conference. Hamilton, New Zealand: Wai-Ora, Wai-Maori, Waikato. http://dx.doi.org/10.4236/oalib.1100804
Henczel, S., & Macauley, P. (2013, 20-23 October,). Professionalism: Exploring the role and responsibility of our professional associations. LIANZA Conference. Hamilton, New Zealand: Wai-Ora, Wai-Maori, Waikato.http://dx.doi.org/10.4236/oalib.1100804
Herzberg, F. (1968). One more time: How do you motivate employees? Boston: Harvard Business Review.
hgcl.kannuruniversity.ac.in
http://library.kau.in/in
https://library.ssus.ac.in
https://www.kvasu.ac.in/
Inceoglu, Ilke., Segers, Jesse., & Bartram, Dave. (2012). Age-related differences in work motivation. Journal of Occupational and Organizational Psychology, 85(2), 300–329.

Retrieved from doi: 10.1111/j.2044-8325.2011.02035.

India, Ministry of Human Resource Development. (1986). National policy on Education-1986(p.15). New Delhi: Department of Education.

India, Ministry of Human Resource Development. (2020). National Educational Policy 2020, Government of India.

Irfan,Mohammed P.P. , Haneefa, Mohammed. , & Shyni, K.G. (2015). Workplace learning among library professionals of university libraries in Kerala. DESIDOC Journal of Library & Information Technology, 35(5), 376-381.

Jana, Lakshmikantha., & Panigrahi, Pijushkanti. (2010). Motivating parameters for betterment of library services: An evaluative study among non-government College Librarians in West Bengal. Annals of Library and Information Studies, 57(1), 12-20.

Janelle R. Thomas., & Holley, R. P. (2012). Management versus repetitive tasks – avoiding "working for the weekend" A crash course in motivating library staff faced with seemingly endless tasks. New Library World, 113(9-10), 462 –473.

Jayaraman, Irudayam & Tamizhchelvan, Mottaiachettiar. (2015). An Empirical Analysis of the relationship between marital status and job satisfaction of Library professionals. International Journal of Applied Engineering Research10(8),21383-21391

Jex, S.M., & Britt, T.W. (2008). Organizational Psychology. Hoboke, New Jersey: John Wiley & Sons, Inc.

Jharia, Babita., & Muley, Anjali. (2014). Role of library in rural development. International Journal of Computer Science and Network (online),152-157.Retrieved from http://ijcsn.org/RDFCORDGE-2014/C-044.pdf

Jones, G. R., & George, J. M. (2008). Contemporary management. Boston: McGraw Hill/Irwin

Kacherki, Umeshareddy., & Konnur, P.V. (2010). Motivational factors for increasing work efficiency among engineering College Librarians: A Comparative study of Maharashtra and Karnataka States. SRELS Journals of Information Management, 47(6), 681-688.

Katna, A.K., & Saibaba, B. (1982). Motivation of librarians. Library Herald, 23(2&3), 99-102.

KAU Handbook 2021, Kerala Agricultural University, Thrissur

Keogh, Patricia. (2013). Motivation for grant writing among academic librarians. New Library World, 114(3-4), 151 - 165. doi:http://dx.doi.org/10.1108/03074801311304069

Kerala State Higher Education Council (2019). Official website of Kerala State Higher Education Council. Retrieved 2019, from the official website of Kerala State Higher Education council. http://www.kshec.kerala.gov.in.

Kishore, Jugal. (1981). Personnel Management in Libraries (pp.111-115). New Delhi: Ess Ess Publications.

Kont, Kate-Riin., & Jantson, Signe. (2014). A study of salary constraints in Estonian University Libraries. Slavic & East European Information Resources, 15(1-2), 60–79.

Lahiri, Ramansu. (1988). The human side of the library: A profile of Manipur. Annals of Library Science and Documentation, 35(2), 84-92.

Lamptey, R.B., Boateng, M.S., & Antwi, I.K (2013). Motivation and performance of librarians in public universities in Ghana. Library Philosophy and Practice (e-journal). Paper 911, 1-26. Retrieved from http://digitalcommons.unl.edu/libphilprac/911

library. mgu.ac.in

library.cusat.ac.in

Liz, Farler., & Broady-Preston, J. (2012). Workplace stress in libraries: A case study. Aslib Proceedings,64(3),225-240.doi.org/10.1108/00012531211244509.

Lynch, Beverly P., & Verdin, Jo Ann. (1983). Job satisfaction in libraries: Relationships of the work itself, age, sex, occupational group, tenure, supervisory level, career commitment, and library department. The Library Quarterly: Information, Community, Policy, 53(4),

434-447.

Madukoma, E., Akpa, V.O., & Okafor, U.N. (2014). Effect of training and motivation on job performance of library personnel of University of Lagos, Lagos State, Nigeria. Open Access Library Journal, 1,1-8. doi: 10.4236/oalib.1100804

Maslow, A.H. (1943). A Theory of human motivation. Psychological Review. 50(4): 370–96. doi:10.1037/h0054346

Miao, C. F., & Evans, K. R. (2007). The impact of salesperson motivation on role perceptions and job performance: A cognitive and affective perspective. Journal of Personal Selling and Sales Management, 27 (1),89-101.

Mitchell, T.R., & Daniels, D. (2003). Motivation. Handbook of Psychology. In W.C. Borman, W.C., Ilgen, D.R. & Klimoski, R J.(ed.) (Vol.12, pp.225-54) Industrial Organizational Psychology. New York: Wiley.

Mohammed, S., Tilahun, M., Kote, M., Mama, M., Tamiru, D., & Rosano, A. (2015). Validation of Health Extension Workers Job Motivation Scale in Gamo-Gofa Zone, Southern Ethiopia: A Cross-Sectional Study. International Scholarly Research Notices, 2015. https://doi.org/10.1155/2015/250610

Navalani. (1990). Job Characteristics of Library and Information work as satisfiers and dissatisfiers-A survey. Annals of Library Science and documentation, 37(3), 85-95.

Nazim, Mohammad. (2016). Knowledge sharing attitude of library professionals: A survey. Journal of Knowledge & Communication Management, 6(2),154-173. doi: 10.5958/2277-7946.2016.00013.9

Nguyen, A. D. (2018). Perceived overqualification and withdrawal among seasonal workers: Would work motivation make a difference? (10746206). Available from ProQuest Dissertations & Theses Global A&I: The Sciences and Engineering Collection. (2042942164). Retrieved from https://search.proquest.com/docview/2042942164?accountid=62114

Omol, Victoria Imabong. (2020). Relationship among Job Satisfaction, Leadership Styles and Career Indecision of Library Personnel in Selected Private Universities in South-West, Nigeria. International Journal of Library Science, 9(2): 40-50.

Onuoha, Uloma Doris., Samuel, Toyosi Daniel., & Ojo, Adebowale Ifeoluwa. (2014). Gender and job satisfaction: A study of library staff in Private University Libraries in Ogun State, Nigeria. Journal of New Media and Mass Communication 1(1), 12-19.

Owolabi, K. A. , Ajiboye, Bosede A. , Bakare, Olubunmi.D. , Bello, Taofeek.O., Omotoso, Akinkunmi. O. , & Adeleke, Olateju. (2013). Librarians' attitude towards monetary and non-monetary incentives in university libraries: A case of selected university libraries in Nigeria. Annals of Library and Information Studies, 60(1), 22-26.

Parasuraman, A, et al. (1988). SERVQUAL: A multiple item scale for measuring consumer perceptions of service quality. Journal of Retailing. 64(1), 12-40.

Payne, R. L., & Pheysey, D.C. (1971). G.G. Stern's organizational climate index: A reconceptualization and application to business organizations. Organizational Behavior and Human Performance, 6, 77–78.

Pinder, C.C. (2008). Work Motivation in organizational behaviour, (2nd ed., p. 587). Newyork: Psychology press.

Plate, Kenneth H., & Stone, Elizabeth W. (1974). Factors affecting librarians' job satisfaction: A report of two studies. The Library Quarterly: Information, Community, Policy, 44(2), 97-110.

Popoola, S.O. (2007). Workplace, biographical and motivation factors affecting organizational commitment of records officers in Nigerian federal universities. African Journal of Library, Archives and Information Science, 17(1), 33-44.

Pors, Niels Ole., & Johannsen, Carl Gustav. (2002). Job satisfaction and motivational strategies among library directors. New Library World, 103(6), 199 – 209.

Prytherch, Ray. (1995). University Library. In Harrod's Librarians' Glossary (8th ed., p.663). England: Gower.

Raj, Sini S., & Devakumar, P.S. (2018). Motivation and job involvement among the employees of General Co-operative Marketing Societies in Kerala, International Journal of Management, IT & Engineering, 8(6), 253-269.

Rajpurohit, Rajpal Singh. (2016). Job satisfaction with regard to social recognition among the library professionals in Medical, Dental and Ayurvedic College of Rajasthan: A study. International Journal of Management, IT & Engineering, 6(12), 125-133.

Ranganathan, S R. (1957). The Five Laws of Library Science (2nd ed.). Bombay: Asia Publishing House.

Rowley, Jennifer. (1996). Motivation of staff in libraries. Library Management, 17(5), 31 – 35.

Sadeghi, S., & Sattari, S. (2018). A Comparative Study of Job Motivation and Work Engagement among Teachers Working in Smart and Ordinary Middle and High Schools in Parsabad. Journal of Ecophysiology and Occupational Health, 18(1-2), 12-17.

Sharma, Sanjeev., & Deolia, Ashish. (2017). Job involvement as a key component of work culture: A study based on survey of the library professionals of the special libraries of Delhi NCR. SSARSC International Journal of Library, Information Networks and Knowledge, 2 (1), 1-17.

SPSS. (2016). Wikipedia. Retrieved from http://en.wikipedia.org/wiki/SPSS.

Sreedhara Menon, A. (1979). Social and Cultural History of Kerala (pp.3-4). New Delhi: Sterling Publishers Pvt. Ltd.

Srivastava, Ashok. (1997). Work Motivation among school librarians in Jaipur: A comparative study of Government and Private Schools. ILA Bulletin, 33(1&2), 11-15.

Srivastava, S. K., & Barmola, K. C. (2011). Role of motivation in higher productivity. Management Insight. VII (1),88-99.

State Planning Board. (2018). Economic Review, 2018. Thiruvananthapuram, Kerala, India. Retrieved from http://www.spb.kerala.gov.in

Susan Mathew, K., Baby, M.D., & Sreerekha Pillai S. (2011). Professional development of academic library professionals in Kerala. Asia-Pacific Conference on Library & Information Education & Practice, 2011. 140-148.

Tella, Adeyinka., Ayeni, C O., & Popoola, S. O. (2007). Work motivation, job satisfaction, and organisational commitment of library personnel in academic and research libraries in Oyo State, Nigeria. Library Philosophy and Practice (e-journal), 1-16. Retrieved from http://digitalcommons.unl.edu/libphilprac/118

Thapisa, A.P.N. (1993). Structures, efficiency and motivation among Librarians. Library Management, 14(5), 20 – 26.

Tietjen, M.A., & Myers, R.M. (1998). Motivation and job satisfaction, Management Decision, 36 (4), 226-231.

Toth, Mate. (2005). Career motivation of Hungarian library school students: a quick survey conducted among the fulltime students of 4 LIS departments. Konyvtari Figyelo, 51(4), 705-733.

Walton, R. (1975). Criteria for quality of work life. In Davis, L.E and Cherns, R.L (Eds.) Quality of work life: Problems, prospects, and the State of the Art (Vol. I, pp.12-15). New York: Free Press.

Williamson, Josephine. (1998). Happily ever after? Motivating staff in times of change. The Serials Librarian: From the Printed Page to the Digital Age, 34(3-4), 361-365.

www.kulib.in

Yahya, MSY., Wan Taib, W.M.A., Che Rusuli, M.S., Tasmin, R., & Hashim, N. (2012). Improvement of job design for Library Assistants. Proceedings of INTED 2012 Conference. 5-7 March 2012, (pp.2507-2514). Spain: Valencia.

Yalokwu, P.O. (2006). Fundamentals of Management (2nd ed., p.396) Lagos: African Centre for Management and Education.

Yaminfirooz, M., Yaminifirouz., & Mousavi, A. (2013). Assessment of organizational climate of libraries at Babol University of Medical Sciences in 2012. Journal of Health

Index